GREAT FEUDS
IN SCIENCE

GREAT FEUDS
IN SCIENCE

*Ten of the Liveliest
Disputes Ever*

Hal Hellman

JOHN WILEY & SONS, INC.
New York · Chichester · Weinheim · Brisbane · Singapore · Toronto

Published by John Wiley & Sons, Inc.
Published simultaneously in Canada.

This publication is designed to provide accurate and authoritative information
in regard to the subject matter covered. It is sold with the understanding that
the publisher is not engaged in rendering professional services. If professional
advice or other expert assistance is required, the services of a competent professional person should be sought.

Library of Congress Cataloging-in-Publication Data

Hellman, Hal
    Great feuds in science : ten of the liveliest disputes ever / by Hal Hellman.
      p. cm.
    Includes bibliographical references and index.
    ISBN 978-1-62045-676-7
    1. Science—History. 2. Scientists—Social aspects—History.
  3. Vendetta—Case studies. I. Title.
    Q125.H5574 1998
    509'.2'2—dc21   97-39824
                                                CIP

10 9 8 7

CONTENTS

To Sheila, Jill, Jennifer, and Scott,
extended nuclear family

ACKNOWLEDGMENTS

Though the final sprint to get this book done occurred in 1996 and 1997, I have been collecting material for it for almost two decades. This included several trips to Europe for some on-site visits, where the hosts and docents were unfailingly pleasant and informative, as at Darwin's home in Downe and Newton's quarters at Cambridge. I am especially grateful to Dr. Franco Pacini, Director of the Osservatorio Astrofisico di Arcetri, Italy, who went out of his way to facilitate my visit to Il Gioiello, where Galileo spent his last, sad years, and who provided some very useful information as well.

The vast majority of my research, however, has been in libraries. Most helpful have been the wonderful collections of old materials at the Burndy Library (Norwalk, Connecticut); the Marine Biological Laboratory (Woods Hole, Massachusetts), where I spent several months at a rented desk in the library; the American Academy in Rome, where I spent a month as research scholar; the Bobst Library at New York University, where I taught for eight years; the New York Public Library in New York; and the newer Science, Industry and Business Library, also in New York. And, finally, I made much use of my own local library in Leonia, New Jersey (luckily part of a countywide library system), from which I and the ever-helpful librarians have been able to reach out and pluck a remarkable number of gems from other libraries in our own county and, when needed, across the country.

Because of the book's wide scope I had to rely, to a large extent, on secondary sources. Among the most useful were books that concentrated entirely or largely on a few of the individual feuds, or that provided information that would have been difficult to track down, such as Rachel H. Westbrook's *John Turberville Needham and His Impact on the French Enlightenment* (Ph.D. Thesis, unpublished, Columbia University, 1972); A. Rupert Hall's *Philosophers at War. The Quarrel between*

Newton and Leibniz (New York: Cambridge University Press, 1980); Elizabeth Noble Shor's *The Fossil Feud between E. D. Cope and O. C. Marsh* (Hicksville, NY: Exposition Press, 1974); and, astonishingly, five separate books on the Mead/Freeman feud.

For concentrated writing time, there is nothing better than the several artists' colonies where I was able to spend some time. These include the Karolyi Institute in Vence, France (now, unhappily, no longer in operation); Mishkenot Sha'anim, in Jerusalem, Israel; and the Fundación Valparaíso, in Mojácar, Spain, where I was able to bring the book to a long-awaited conclusion.

Over the years I have pestered a great number of colleagues with questions. Some of these scholars I knew, some I did not. In almost all cases, they were helpful. Too many to name, I thank them nonetheless. One whom I would like to single out, however, is Shirley A. Roe. It was her lecture, "Voltaire versus Needham: Spontaneous Generation and the Nature of Miracles," at the New York Academy of Sciences on December 2, 1981, that got me started in the first place.

Several colleagues were kind enough to read and comment on portions of the manuscript. These include Morton Klass, Professor Emeritus of Anthropology at Barnard College and Columbia University; Samuel I. Mintz, Professor Emeritus of English at the City University of New York; Walter Pitman at the Lamont-Doherty Geological Observatory in Palisades, New York; Phyllis Dain, Professor Emeritus of Library Science at Columbia University; Norman Dain, Professor Emeritus of History at Rutgers University; John R. Cole, anthropologist and board member of the National Center for Science Education (former president); and Dr. Harold L. Burstyn, an escapee from the history-of-science field, who is now a patent lawyer in Syracuse, New York.

Thanks also to my agent Faith Hamlin for her sound marketing and psychological insights; to my editor Emily Loose for her input; and to John Simko, who saw the book through to the end.

Finally, the support, encouragement, and prodding from my wife, Sheila, who read every word at least a couple of times, was most important of all.

INTRODUCTION

James Galway, the Irish flutist, was scheduled to perform with the London Chamber Orchestra. Because one of the pieces, composed by Carl Stamitz, was not well known, he felt he would be more comfortable conducting that piece himself. It was then decided that Galway would conduct the entire concert.

"I got the tempo of the first piece very clear in my head," he notes. "Then I gave the downbeat and we began. When I saw the expression of horror on the players' faces, I knew we had a disaster on our hands. We just stopped dead. What else could we do? I had neglected to look at the program. We were supposed to start with a Vivaldi, not the Stamitz."

A conductor's worst nightmare—but here comes a surprise. "We just cracked up," he adds. "And when the audience realized what had happened, they joined in. I think they like to see something go wrong sometimes."[1]

Unfortunately, when the media report that something has gone wrong in science, it is likely to be a calamity of some sort: the release of a deadly chemical—as in earlier incidents involving methyl mercury, dioxin, and PCBs—or a nuclear accident such as those of Three Mile Island and Chernobyl. Though these disasters should probably be classified as technological rather than scientific, reporters rarely make this distinction.

News reports of errors in science, rather than technology, seldom appear. As a result, the public rarely sees the various wrong paths down which scientists often tread. Even if an incorrect scientific idea is reported, no one knows it is incorrect. By the time the correct idea is developed, it is brought forth as a new breakthrough, and the old one is simply forgotten. Even in scientific journals, reports of negative results rarely make it into print, in spite of the fact that they can be very helpful to those working in the field.

Part of the problem lies in how science is taught—namely, as a kind of grand march. Almost any science textbook presents the material as a logical series of chapters. Like a juggernaut, the text grinds through the science to be covered, never diverting from its path to show what a struggle it was to bring the ideas to fruition.

After all, the facts, even the theories, are history. It is the *process* that is the living science; that's what makes the activity exciting to those who practice it. Nonscientists really don't understand this. They are equally in the dark when it comes to the practitioners of science, whom they think of in the same way as the work the scientists are carrying out: cold, unfeeling—in short, inhuman.

Often, however, the process of scientific discovery is charged with emotion. When introducing a new idea, a scientist is likely to be stepping on the theories of others. Holders of an earlier idea may not give it up gladly.

How does the loser feel when he or she sees a cherished theory being overturned, perhaps even sees immortality slipping away? When the loser goes down fighting, we have one kind of scientific feud, as illustrated in the quarter-century-long battle between Thomas Hobbes and the British mathematician John Wallis (see Chapter 2). One of Hobbes's problems was that he was so taken with geometry as to be wholly blind to the capabilities of algebra. Thus, he could in all honesty say of Wallis's ingenious algebraic method that it was "so covered over with the scab of symbols" that he had no patience to deal with it, and that it looked "as if a hen had been scraping there."

Another source of scientific disagreement has to do with questions of priority; such questions may arise when two or more scientists come up with the same idea at about the same time. While simultaneous discovery in science and mathematics seems amazing, it is actually fairly common; examples include Newton and Leibniz (calculus), Faraday and Henry (electromagnetic induction), Adams and Leverrier (discovery of Neptune), Darwin and Wallace (theory of evolution), and Heisenberg and Schrödinger (quantum mechanics).

Certainly, one of the major drives in science is just the pleasure of finding things out, learning something new about the world around us. If scientists were saints, they might be satisfied with that; for the most part, they are not driven by monetary gain. If they discover something, however, they generally want the world to know it. Visions of a Nobel Prize may dance before them.

All such cases, then, offer the opportunity for priority conflicts, and some ferocious battles have resulted. For this book, I have chosen what I consider to be the most dramatic examples, and we'll see a variety of responses to the challenges.

We'll also see, however, that even defining a discovery, or deciding when it occurred, is not alway easy. Sometimes the question has been resolved in a polite, courteous fashion—as in the case of evolution, involving Charles Darwin and Alfred Russel Wallace—so we can safely ignore that story.

Other cases, however, have involved bitter recriminations. A classic example is the battle between Isaac Newton and Gottfried Wilhelm Leibniz (see Chapter 3). Though Newton was a difficult man and battled with others as well for a variety of reasons, he was troubled by all these conflicts. These feuds may even have soured his taste for the world of active science.

Why do some controversies resolve satisfactorily, while others seem to continue on and on? In the latter case, the science itself may be recalcitrant, just plain slow to develop. As a result, competing ideas go back and forth. More often, there is some subtle or not-so-subtle question of beliefs or values that underlies the whole debate. Many stories are of this kind and provide models that can help us understand conflicts that still rage today, as in the case of creationism (Darwin, Chapter 5) and also in that fiery question, When does the lump of developing tissue in a woman's uterus become a human being? (See Voltaire versus Needham, The Generation Controversy, Chapter 4.) In this decade-long feud, Voltaire took it upon himself to expose Needham, a well-known English naturalist, as a "dangerous biological thinker." Voltaire's methods included one that is not unknown in our own day: He suggested that Needham was a homosexual. Needham, in answer, scornfully referred to "so-called sages" who rigorously profess, but do not practice, celibacy—a shot at Voltaire's several love affairs, the latest of which was with Voltaire's own niece.

Also, although Donald Johanson and Richard Leakey are no longer slugging it out in print and in person (Chapter 9), the question of human origins remains as contentious and tantalizing as ever.

With Johanson and Leakey, the positions of challenger and challengee reversed at one point. That surely was not the case when Derek Freeman, a little-known Australian professor of anthropology, went after that American icon, Margaret Mead, for Mead was already dead. The resulting fracas (Chapter 10) drew in a remarkable collection of defenders and attackers. Although the exchange is still going strong, the truth is that Mead's reputation has suffered.

Another, more positive, outcome from a feud is seen in the current excitement about dinosaurs, which owes much to the bitter rivalry between two great fossil hunters of the nineteenth century. The competition between Edward Drinker Cope and Othniel Charles Marsh (see Chapter 7) became legendary and included every form of duplicity and chicanery possible. Nonetheless, it made *dinosaur* a household word and kindled public interest in paleontology. This interest led the public to greater support of museums and expeditions, and this, in turn, brought about more discoveries.

In some cases, new ideas in science were threatening to the beliefs of the public, as well as to other scientists. The theory of evolution is a good example, and among the attackers were many laypersons. (In fact, the word *scientist* did not even exist prior to 1840; it was coined in that year by the English scholar William Whewell.) One of the most famous debates in the history of science was that between Darwin's defending bulldog Thomas Henry Huxley, who was a scientist, and Bishop Samuel Wilberforce, who was not (Chapter 5).

Lord Kelvin, a contemporary of Darwin's, had no need of defenders. He was so widely respected that his ideas on the age of the Earth, though quite wrong, held sway for an astonishing 60 years (Chapter 6).

Alfred Wegener, on the other hand, had no such reputation. As a result, he had to fight long and hard before his remarkable insights into continental drift were recognized (Chapter 8).

Here, then, is a book on great feuds in science. Organized in roughly chronological order, the chapters make up an ongoing conversation that spans the whole history of modern science, including a bit of the mathematics that is so basic to the discipline. It is a dialectic; by considering the introduction and growth of a new idea and at the same time the sea out of which it sprang, we may come away with a better appreciation for what has been accomplished.

In the same way that political history helps heads of state interpret today's events, these short dramatic episodes tell us about science as both a human enterprise and an organized activity.

In other words, I want to show that scientists are susceptible to human emotions; that they are influenced by pride, greed, belligerence, jealousy, and ambition, as well as religious and national feelings; that they are subject to the same frustrations, blindness, and petty emotions as the rest of us; that they are in truth fully human. As a result, this is a history of the losers as well as the winners.

We begin with the case of Pope Urban VIII versus Galileo. Some writers argue that this feud was the start of a still-existing schism between science and religion. At least one writer, History of Science Professor William Provine, argues that the split came later and arose out of the evolution controversy.[2] Perhaps, as some maintain, there is no schism at all? We could argue about that.

CHAPTER 1

Urban VIII versus Galileo

An Unequal Contest

Entering Saint Peter's Basilica in Rome is like walking into the Grand Canyon. One experiences the same feelings of grandeur, of awesome majesty. Both are on such a grand scale that a mere human feels insignificant—which, in the case of Saint Peter's, is exactly what its creators sought.

The largest religious edifice in the world, the basilica is two football fields long, covers 4 acres, and can hold up to 50,000 people. In one of its vast mosaics, Saint Mark's pen is almost 5 feet long! The basilica took more than a century to create and build, and practically all the great architects and artists of the late fifteenth, the sixteenth, and the early seventeenth centuries—including Michelangelo, Raphael, Bernini, Sangallo, and Bramante—were involved in its design. Marble, bronze, gilt, and soaring space combine to create an overpowering experience.

Only slowly do other aspects of the remarkable structure begin to resolve into intelligible details. One such detail is the great bronze baldacchino—four huge spiraling columns supporting a magnificent bronze canopy over the tomb of Saint Peter. Rising to the height of the Palazzo Farnese not far away, it dominates the center of the basilica.

On approaching the baldacchino, visitors see yet another set of details emerge. At the base of the columns are some curious oval bas-reliefs, sculpted in marble, depicting three bees in formation. This pattern is the coat of arms of the Barberini family; it appears no less than eight times around the base of the baldacchino, and around the top as well.

1

The Barberinis can be traced back to eleventh-century Florence. By the sixteenth century, the family had amassed a fortune and wielded great influence. In 1623, Maffeo Barberini, then age 55 and a cardinal, was elected Pope Urban VIII, thereby adding to the family's financial and political power the might of the Roman Catholic Church. In due course, Urban spread some of this additional influence around. He made a brother and two nephews cardinals. To a third nephew he gave the principality of Palestrina.

Urban didn't begin Saint Peter's, but it was brought to completion during his reign, and his mark is seen not only on the baldacchino but also throughout the huge structure. Most obvious is a stupendous bronze sculpture of his holiness, with his right arm upraised in benediction, or perhaps warning. He is accompanied by two marble figures, representing Charity and Justice. A great marble tablet over the entrance to the basilica proclaims Urban's importance to the building.

His bees are also seen swarming among the bay leaves decorating the baldacchino's columns and elsewhere in Rome as well—in the great Palazzo Barberini, which now houses the Rome National Gallery; and in the delightful Fontana del Api, the fountain of the bees.

Some scholars say that the Barberini bees symbolize the original name of the pope's family, Tafani, meaning "gadfly"; others maintain that the bees are the symbol of divine providence; still others say they represent industry and productivity, and Urban was indeed busy rebuilding and beautifying the city. Nonetheless, one is hard put not to think instead of the bee's sting. There was, for example, the case of Marco Antonio de Dominis, a relapsed heretic who died in prison before his trial, but whose body and works were burned under Urban's rule in 1624.

Such was the power facing Galileo Galilei when he and Urban squared off in the early years of the seventeenth century. By a great irony, while the beautiful baldacchino was being dedicated in Saint Peter's, where Charity and Justice accompany Urban, he was clamping down hard on Galileo, who had dared to challenge his might.

On June 22, 1633, Galileo Galilei was put on trial at Inquisition headquarters in Rome. All the magnificent power of the

Roman Catholic Church seemed arrayed against him, this old man of 69; in his defense he referred to his "pitiable state of bodily indisposition." Under threat of torture, imprisonment, and even burning at the stake, he was forced, on his knees, to "abjure, curse and detest" a lifetime of brilliant and dedicated thought and labor. Charged with "vehement suspicion of heresy" he had to renounce, "with sincere heart and unfeigned faith," his belief that the Sun, not Earth, is the center of the universe, and that Earth moves around the Sun and not vice versa.

Because he was willing to do this—at least verbally—the more serious of Urban's threats remained only that. As one of his punishments, for example, he was to recite the Seven Penitential Psalms once a week for three years. He was, however, placed under house arrest for the rest of his life. Finally, his book *Dialogue on the Great World Systems, Ptolemaic and Copernican* (1632), which lay at the heart of the trial, was prohibited. That is, it was added to the index of banned books, *Index librorum prohibitorum,* maintained by the Inquisition of the Catholic Church.

The Battleground

Ten cardinals sat in judgment of Galileo. Urban VIII was not present in person, but he was there in spirit, for his personal feelings of anger and frustration were the driving force behind these extraordinary proceedings. In fact, of the 10 cardinals present at the trial, only 7 signed the final decree, almost surely indicating a lack of unanimity among them.

It may be that the pope's ignition temperature was lower than normal at the time. While Urban's conflict with Galileo was to loom large in the history of science, it posed only one more in a large pack of troubles bearing down on the Holy Father, for the Thirty Years War was raging during his pontificate, with Catholic and Protestant armies locked in battle in many areas of Europe. To forestall possible invasion, he was deeply engaged in building up the fortifications for Castel Sant' Angelo, the papal fortress, and in other defensive measures as well.

At the same time, he was beset by reverses in many areas. He had been outwitted by Cardinal Richelieu in a complicated power play; he saw the huge papal domain being forced back into the Hapsburg Empire; and, finally, he recognized just how seriously Galileo's new science challenged established church doctrine. The handwriting on the wall was only too clear. Worse, as Galileo put it, the book of nature was written in the language of mathematics, not in biblical verse.

Urban had been elected pope in 1623 at age 55. Until then, he had been Cardinal Barberini, to all accounts a warm, compassionate, intelligent human being, one of the few with whom Galileo felt he could discuss his work intelligently. The political reverses, however, along with the demands of high office, or perhaps the power that came with it, had changed Urban from a warm and compassionate man to one of quick temper and great suspicion. One of Urban's great suspicions was that he had been tricked and betrayed by Galileo.

Galileo had followed strict protocol; had had his book scrutinized by official church censors; had received the church's official imprimatur—and had clearly fooled all the officials into thinking that his ideas were only being presented as hypotheses, which made them acceptable to the church. He had almost gotten away with publishing a heretical work without provoking Urban's wrath.

What made him think he could get away with it? Prior to publication of the *Dialogue,* the pope had counted himself one of Galileo's good friends and admirers. During one of Galileo's visits to the holy city, shortly after Urban's election, the famous scientist was granted six audiences; each lasted more than an hour, an extraordinary allocation of the pope's time. In fact, it was largely because of Urban's election to this high office that Galileo began to think he could safely write the *Dialogue.*

Both men had been born and raised in Florence, and both had attended the University of Pisa, where Galileo studied medicine and Urban had taken a law degree. As Cardinal Barberini, Urban had even interceded in Galileo's behalf during an earlier confrontation with the Holy Office. At that time, 1616, Galileo had been warned that his support of the concept of a sun-centered universe could get him into trouble. He was then told that he could consider the concept as long as he thought of it as a hypothetical idea. He was not to present it as a reality, however, and should not even think of it in that way.

In 1632, some 16 years after that foreboding event, Galileo was a widely known and respected scientist, as well as the official astronomer and philosopher at the court of the Grand Duke of Tuscany. There was, in all probability, some arrogance involved in Galileo's decision to publish his *Dialogue.*

Of more importance, however, were his feelings about religion, for this was no scoffing atheist, nor even an angry escapee from religion. He had attended Catholic school; both of his daughters had become nuns; and, most important, he considered himself a loyal son of the Holy Mother Church. He felt, in other words, that he was trying to save, not to hurt, the church. He was trying, desperately, to prevent the church from putting itself into the position of defending a doctrine that was, in his mind, subject to disproof.

Evidence of his amazing, and continued, loyalty is seen in a letter he wrote in 1640, seven years after the trial: Blind and still under house arrest, and after having been forced to curse and revile the *Dialogue* for several years, he commented (in a letter to Fortunio Liceti) on the question of whether the universe is finite or infinite. He concluded that "only Holy Writ and divine revelation can give an answer to our reverent demands"[1]: still a believer, and hardly a wild-eyed revolutionary.

Like the Italian philosopher Giordano Bruno before him, he did incline to the idea of an infinite universe, but he refused to speculate on its implications, one of which, clearly, is multiple inhabited worlds. In the eyes of the church, both ideas were clearly heretical. Bruno, who had not been as wary as Galileo, advocated the ideas in no uncertain terms and as a result was hauled up before the Inquisition; for refusing to recant, he was burned at the stake in 1600. Galileo was well aware of the disastrous outcome of Bruno's contest with the Holy Office.

Yet he continued to develop his ideas, and he continued to be attacked because of them. In earlier years, he had come into conflict mainly with his teaching contemporaries at Pisa and Padua because of his attacks on Aristotelian physics. Though he began to support the idea of a sun-centered universe as far back as the late 1500s, it was not until the years 1612–1614 that his defense of the system brought him into conflict with the church.

Even Galileo was surprised by the intensity of the pope's reaction. In fact, one needed only to bring up Galileo's name, as some of his friends did in hopes of softening the pope's anger, and Urban would explode in fury. At one point before the trial, the Tuscan ambassador to Rome, a good friend of Galileo's, merely entered the pope's chamber and was met by an angry blast: "Your Galileo has ventured to meddle with things that he ought not to, and with the most important and dangerous subjects that can be stirred up these days."[2]

The Two Chief World Systems

As everyone knew, Nicolaus Copernicus had proposed the *heliocentric* (sun-centered) system almost a century earlier, in his own book, published in 1543. Copernicus himself, a canon in the Polish Catholic Church, had recognized the possibility of trouble and had delayed publication for many years. In a scene that could have been written by a Hollywood screenwriter, the first copy of Copernicus's book came off the press and was put into his hands even as he lay on his deathbed.

Or so legend has it. More certain is that he had overestimated by far the impact of his writing, for it turned out to be one of the major unread works of all time. As long as the doctrine lay shrouded in Latin, just another long-winded academic treatise that few read or cared about, it could be safely ignored by the Catholic Church. Martin Luther, however, smelled something; he called Copernicus the "new astrologer" and predicted, "The fool will overturn the whole art of astronomy."[3] But the book never even made the index, a sure sign of the work's impotence—at least not until 1616, when Galileo's support of the doctrine forced the church to recognize the fertility of Copernicus's idea.

To help us understand the new system, it will be useful to take a quick look at the old one. Watch the sky carefully over a period of time; what do you see? Clearly, the heavenly bodies are all revolving around Earth. But the movement is by no means a simple, regular motion. The planets, especially, have their own timetables and do not move in simple, steady paths. Some of them even seem to double back on their paths now and then.

About A.D. 150, Ptolemy of Alexandria, astronomer and geographer, put together an astronomical system to explain his nighttime observations. Ptolemy's solution was a system in which Earth was at rest in the center of the universe, with the Moon, Sun, planets, and stars revolving around it, all embedded in a system of concentric, crystal spheres.

The advantage of Ptolemy's system was that it worked. That is, it enabled astronomers to predict with some accuracy the motions of the heavenly bodies. For the calculations, Ptolemy assumed that all the heavenly bodies moved in circular paths. To help these match up with observed activity, which is much more complicated, he added a set of additional smaller circular orbits, called epicycles. The result was a very complex geometry, but it was the best there was. It even served as a basis for establishing a set of planetary tables, which predicted the positions of the planets at various times.

In the mid–thirteenth century, the Spanish king, Alfonso X, sponsored a revision of the planetary tables, to bring them in line with later observations. During the long, tedious preparation of the tables, Alfonso, who was footing the bill, commented that if the Lord had asked his advice, he would have recommended something simpler.

Copernicus's idea stood Ptolemy's on its head. Copernicus, like Alfonso, considered the Ptolemaic system too complex. He hypothesized as follows: Suppose that the Sun is at rest and the Earth has a twofold motion—namely, that it rotates once a day on its axis, and that it revolves around the Sun once a year. Simple as that.

Copernicus was not the first to advance this heliocentric idea. It had been proposed much earlier by several of those amazing Greeks, including Aristarchus of Samos, around 260 B.C. He, like Galileo, was denounced for impiety, but the denunciation apparently did him no harm. Aristarchus could advance no proof for the heliocentric idea, however, and it went into hibernation.

Ptolemy's system was really the first that was thorough enough to be able to deal with the observed mass of celestial motions. Of course it matched what people "saw with their own eyes." Later, Ptolemy's description of the universe became entrenched in the Catholic Church's teachings, largely through the work of Saint Thomas Aquinas, a thirteenth-century theologian and philosopher. The centrality of mankind, for example, as an important part of Christian teaching, clearly meshes nicely with an Earth-centered (geocentric) cosmology.

The Christian idea of heaven and hell also meshed beautifully with the geocentric system, which saw the heavenly bodies as not only perfect, but also immutable. In other words, everything in heaven is eternal and incorruptible, whereas growth and, especially, degeneration and decay are restricted to Earth, punishment for the sins of our biblical forebears.

Astronomical references are not difficult to find in the Bible. From Psalm 93: "Yea, the world is established; *it shall never be moved.*" In Psalm 19: "The heavens are telling the Glory of God; and the firmament proclaims his handiwork. . . . In [the heavens] he has set a tabernacle for the sun, which comes forth like a bridegroom leaving his chamber, and like a strong man runs its course with joy. *Its rising is from the end of the heavens, and its circuit to the end of them*" (italics added). What could be clearer? Also, how could Joshua have made the Sun stand still if it had not been moving in the first place?

These lines are clear statements of the ancients' astronomical beliefs. But are they enough to have made Copernicus hesitate, and to create all kinds of problems for Galileo? Today, no. In the fifteenth and sixteenth centuries, they definitely did.

It is hard for us to realize, in our secular era, how pervasive was the influence of the Catholic Church at that time. Every event was a sign of God's anger or pleasure—or of Satan's. Comets were harbingers of disaster. Though the Italian universities were not under direct control of the church, all the teachers were imbued with religious doctrine, and most of them were clerics. (Among the very few exceptions was the University at Padua, where Galileo taught and worked from 1592 to 1610.) Even medicine was to a great extent an amalgam of religion, superstition, and faith.

In such an atmosphere, the heliocentric universe really was a jarring concept. It was the implication, more than the theory itself, that was so jarring. As brave a turnabout as the Copernican theory was, it did not offer any significant gain in simplicity; nor did it offer any gain in accuracy. Copernicus was still hung up on the idea that the orbits of the heavenly bodies must be circular—because circular motion was the most "perfect" type of motion. This fixation on circular orbits forced him to move the center of the system away from the center of the Sun, where it belongs, thus depriving his system of the basic simplicity that would otherwise have been its major advantage.

Copernicus's beliefs differed from contemporary beliefs in other ways as well. For instance, what made the heavenly bodies move across the sky? Angels, said Aquinas. Oh, no, said Copernicus, it is in the nature of perfect circles to rotate forever.[4] The basic reason for his belief in his heliocentric theory is also instructive—namely, that there can be "no better place than the center for the lamp that illuminates the whole universe."[5]

It remained for Johannes Kepler, a German astronomer/physicist/mathematician, to move the heliocentric juggernaut onto the right track, mainly through his discovery that the planetary orbits were elliptical, not circular. Yet Kepler, like Copernicus, was apparently led to support the heliocentric idea through his own kind of Sun worship.

Strangely, although Galileo and Kepler were contemporaries, and had even corresponded, and although Kepler was one of the few other major scientists who supported the heliocentric idea, Galileo never made use of his work. Galileo, too, clung to circular orbits—an indication of how difficult it is to break out of an old mold.

Evidence

There were, in any case, objections to the heliocentric theory that still had to be answered. After many years of argument,

Galileo finally recognized that something more substantive was needed. Somehow, he felt, he had to demonstrate the truth of his arguments, but he found no existing evidence that he could use.

A significant part of the evidence Galileo put forth, then, was his own, based on his own telescopic observations, and conducted using a telescope that he designed and built himself. In answer to the Scholastics' objection that a body cannot have two motions at once, he produced Jupiter's satellites, which clearly were revolving around Jupiter while Jupiter revolved around Earth (or the Sun—it doesn't really matter to the argument). Dealing with the traditional claim that the heavenly bodies are perfect, Galileo showed that the Sun has spots and that the Moon is not smooth but has mountains. As for the Scholastics' objection that the Copernican doctrine required Venus to show phases, which had not hitherto been seen, Galileo claimed that his observations showed Venutian phases, too.

It must be kept in mind, however, that these sightings were being made, mainly in the years 1609 and 1610, through very primitive telescopes. It took a practiced eye to make any sense of them, and many of Galileo's contemporaries who did look saw nothing but jiggling blurs of light. Others simply refused to look. One of the scientists who boycotted the telescope was Professor Giulo Libri. Upon this good man's death a few months later, Galileo suggested that although Libri would not look at the celestial objects while on Earth, perhaps he would take a view of them on his way to heaven.[6]

Well aware of the power of the church, Galileo knew that its blessing would be necessary if his telescopic sightings and his advocacy of heliocentrism were to amount to anything. In 1611, he set out for Rome on a very special kind of pilgrimage. Remember, however, that Galileo was no ordinary supplicant for a favor. M. Berti, a nineteenth-century scholar who was among the first to be permitted to examine the Vatican files in later years, wrote of him:

> Would we form an idea of how Galileo was appreciated and courted at Rome, we must figure him to ourselves in the vigor of life, at the age of forty-seven, with ample forehead, grave countenance, expressive of profound thought, fine figure and very

distinguished manners, clear, elegant, and pleasing, and at times imaginative and vivid in discourse. The letters of the time superabound in his praise. Cardinals, patricians, and other persons in authority vied with each other for the honor of having him in their houses, and hearing him discourse.[7]

Prior to this time, Galileo's opponents had been academics, virtually all of whom were mired in the swamp of Aristotelian science. But Galileo was a powerful and sometimes sarcastic arguer and had made many enemies among his peers. The result was that they began taking potshots at him behind his back. Then, when all else failed, says Giorgio de Santillana, one of Galileo's major biographers, his enemies decided to throw the church at him.[8]

Even if they hadn't instigated trouble, however, it is possible that his telescopic observations alone would have had the same effect. His *Letters on the Solar Spots,* published in 1613, offered the first published statement that the heliocentric theory is the only one that fit his telescopic observations. He concluded triumphantly, "And perhaps this planet [Saturn] also, no less than horned Venus, harmonizes admirably with the great Copernican system, to the universal revelation of which doctrine propitious breezes are now seen to be directed toward us, leaving little fear of clouds or crosswinds."[9]

Trouble was already brewing in the Catholic Church, however. Father Lorini maintained that "the doctrine of Ipernicus, or whatever his name is," was against Holy Scripture.[10] The following year saw the church's first open attack on Galileo's position. Tommaso Caccini, a young Dominican hothead, struck out at the new astronomy from the pulpit of the Church of Santa Maria Novella in Florence. Denouncing Galileists, and all mathematicians along with them, he reportedly used as his text a passage from the Book of Acts: "Ye men of Galilee, why stand ye gazing up into heaven?"[11] Though the line can be taken as a humorous pun, there was little humor in the furious sermon preached by Caccini.

By 1616, Galileo was being warned by Cardinal Bellarmine that he was treading on dangerous ground. The church's position was made very clear in a letter written by Bellarmine at that time. Commenting on a work by the Carmelite priest Paolo

Antonio Foscarini, which supported the Copernican system, Bellarmine pointed out, "I say that if there were a true demonstration that the sun was in the center of the universe . . . then it would be necessary to use careful consideration in explaining the Scriptures that seemed contrary. . . . But I do not think there has been any such demonstration."[12]

Bellarmine was right. All the evidence that Galileo had offered, specifically the telescopic observations, showed that Earth *could* be revolving around the Sun, but in no way proved that it *was*. The point is that if such demonstration were available, it would obviously tear a significant portion of church doctrine to shreds. Until then, it was far better for the church to maintain the status quo in hopes that the whole disturbing situation would just dry up and blow away after a while.

Possibly, if Galileo had not thought of doing the *Dialogue*, it would have, at least for a while. He saw what had to be done, however, and he went ahead and did it. Why did his book stir up a hornet's nest, when Copernicus's treatise did not? The main problem with Copernicus's work, as I suggested earlier, was that it was poorly packaged. Galileo's *Dialogue* was quite another matter. True, it was not simple—but it was clever, lively, and eminently readable.

There is an interesting sidelight here, which further illuminates Galileo's situation. In the heyday of the Roman Empire, intellectual discourse and writing were done in Greek, and Latin was the vernacular. By Copernicus's and Galileo's day, however, intellectuals wrote their scholarly works in Latin— mainly because so many of the scholars were associated with the Roman Catholic Church—while the common language was Italian. Galileo wrote his *Dialogue* in Italian, which meant that it could be, and was, widely read and discussed. In contrast with Copernicus's *De Revolutionibus Orbium Coelestium*, then, Galileo's *Dialogue* was an overnight hit—and that the church could not ignore.

The Dialogue

Galileo's *Dialogue* has several English translations, all of which give some of the flavor he tried to impart. They also preserve his own construction, a series of conversations carried out over four days. There are three participants: Salviati, Sagredo, and Simplicio. Salviati, named for an old friend of Galileo's who had died in 1614, speaks for Galileo. It was at Salviati's magnificent villa overlooking the Arno that Galileo, in 1612, carried out his observations on sunspots. Salviati also shared Galileo's enjoyment of burlesque poetry and low comedy.

Sagredo, named in memory of another deceased friend of Galileo, is the intelligent, impartial moderator, a person of high rank and a man of the world. In his early years, Galileo, though serious about his work, was not averse to having a good time, and there are reports of wild parties held on Sagredo's estate on the Brenta River.

The book's third discussant is Simplicio, a composite of all the opponents Galileo had fought along the long road he had traveled. Galileo's technique is first to build his opponents' arguments through Simplicio, adding some of his own, which his opponents never even thought of, and then to demolish those assertions with powerful arguments and, often, devastating satire.

Simplicio, for instance, reflects a common belief of the time that the Sun, Moon, and stars, "which are ordained for no other use but to serve the Earth, need no other qualities for attaining that end save only those of light and motion."

"What's this?" argues Sagredo. "Will you affirm that Nature has produced and designed so many vast perfect and noble celestial bodies, invariant, eternal, and divine, to no other use but to serve this changeable, transitory, and mortal earth? To serve that which you call the dregs of the universe and sink of all uncleanliness?"

A clean cut. Then, turning the knife, Sagredo adds, "I cannot understand how the application of the Sun and Moon to the Earth to effect change should be any different from laying a marble statue in the chamber of the bride and from that conjunction to expect children."[13]

Regarding the heavy reliance of his opponents on classical texts, particularly Aristotle, Sagredo chides, "But, good Simplicio, this reaching the desired conclusion by connecting several small abstracts which you and other egregious philosophers easily find scattered throughout the texts of Aristotle I could

do as well by the verses of Virgil or Ovid, composing patch-
works of passages which explain all the affairs of men and secrets of Nature. But why do I talk of Virgil or any other poet? I
have a little book much shorter than Aristotle or Ovid, in which
are contained all the sciences, and with very little study one
may gather out of it a most perfect system, and this is the
alphabet."[14]

Does it seem that Galileo is firing in all directions at once? It
does indeed—and his *Dialogue* came out to be some 500 pages
long as a result. But there was good reason. Much as he would
have liked to deal directly with the cosmological question—
Ptolemaic versus Copernican systems—he could not. For the
Ptolemaic theory is an integral part of a complex, complete system: science, philosophy, and religion all rolled into one.

Ptolemy, for example, had written,

> Mortal though I be, yea ephemeral, if but a moment
>   I gaze up at the night's starry domain of heaven,
> Then no longer on earth I stand; I touch the Creator,
>   And my lively spirit drinketh immortality.[15]

Science? Religion? Philosophy? Astrology? Poetry? Before
Galileo could get to the cosmological arguments, then, he had
to take apart a huge and powerful—albeit ungainly—edifice,
brick by brick, idea by idea—and that's what he did. Earlier in
life, he had referred to the project as his "Immense Design."
The description is apt.

But all of his arguments, he knew, would avail naught without
evidence. In fact, the early part of his *Dialogue* is really just a
softening-up operation for what Galileo feels are his devastating
blows—the evidence. Toward the end of the book, Salviati has
just explained a connection between Earth's motion and its
tides. To Galileo, this is the clincher: Earth's waters move. That
much is known. Through a long series of arguments, developed
slowly and logically, he shows that this movement of waters is
evidence that Earth does indeed move. Proof.

Sagredo breathes in wonderment: "If you had told us no
more, this alone, in my judgement, so far exceeds the vanities

introduced by so many others that my mere looking on them nauseates me, and I very much wonder that among men of high intelligence . . . not one has ever considered the incompatibility that is between the motion of the water contained and the immobility of the vessel containing it."[16]

Ironically, Galileo also takes a potshot at Kepler, who had suggested that the tides were somehow caused by something in the heavenly bodies. Kepler thought, however, that this heavenly cause was magnetism. In the *Dialogue,* Salviati accuses Kepler of having "given his ear and assent to the Moon's predominance over the water and to occult properties and suchlike trifles."[17] This sort of action at a distance seems to Galileo an example of Kepler's mystical bent.

It was not until much later that Kepler's inspired guess was borne out, for the tides are indeed caused by the Moon's and, to a lesser extent, the Sun's gravitational (though not magnetic) pull. They are *not* caused by the motion of Earth.[18] This is a good example of Galileo's power with words, for even when he was wrong, he was convincing.

A Grave Mistake

Clearly, in order to convince his readers, Galileo had to make his arguments solid and powerful. To make them obvious, and perhaps to vent some spleen, he used Simplicio as a foil. But the more foolish Simplicio's arguments are, the more clear is Galileo's real objective. He resolved to take this chance—and through most of the book, it works.

At the end, however—carried away perhaps by an excess of zeal, and secure in his conviction that he had found a way to let out his feelings without personal danger—he lets Simplicio sum up the Catholic Church's position concerning the impossibility of obtaining true knowledge of the physical world. Simplicio says that if God had wanted to make Earth's waters move in some other way than by making Earth move, he certainly could have done so. "Upon which I forthwith conclude that, this being granted, it would be an extravagant boldness for anyone to limit and confine the Divine power and wisdom to one particular conjecture of his own."[19] The "particular conjecture" to which Simplicio is referring, of course, is the Copernican system.

Simplicio's closing statement doesn't sound very explosive, does it? It seems likely that Galileo felt the same way. Yet Galileo's enemies were later able to convince Urban that if the statement came from Simplicio's mouth, Galileo's intent must have been to make fun of it, and, worse, of Urban himself. Now, Galileo was strong-minded, but he was not stupid. The problem was that Simplicio's assertion had been a standard argument of the pope's, and Galileo had been directed by the censors to include it in the book. Clearly—in Galileo's head anyway—the argument had to come from Simplicio. Conceivably, he even forgot that the argument had been Urban's.

In any case, when Urban saw the result, he was furious—furious and unforgiving. Even after Galileo's death in 1642, Urban refused to relent. The Grand Duke of Tuscany, who had been Galileo's patron for many years, wanted to hold a suitable public funeral for Galileo and to erect a monument over Galileo's grave at the Church of Santa Croce in Florence. Urban warned the Grand Duke that he would consider such action a direct insult. The result was that the remains of one of the great scientists of all time were quietly hidden away in the basement of the church bell tower for almost a century.

Finally, permission was given for Galileo's remains to be interred under a large monument at the entrance to the church, where they lie today. Nearby are the tombs of two other famous Florentines: Michelangelo and Machiavelli. As for the *Dialogue,* the church did not officially release it from the index until 1822—which is not to say that it was not distributed until then. Copies had been smuggled out to other European countries, had been translated into Latin, and had been widely discussed among non-Italian scholars.

Some historians suggest that had Galileo stayed a professor at Padua, in the independent republic of Venice, instead of entering the Grand Duke's service in 1610, he would have been far better off. Would science have been better off? That's harder to answer. Had the trial not taken place, Galileo would no doubt have continued to support and champion the Copernican theory. Because he was forbidden to do so, he turned his attention to writing a book that was to prove much more important to basic science than his *Dialogue* had ever been. This was his *Dialogues Concerning Two New Sciences* (1638)—the summation and distillation of all his earlier work on mechanics. The later book, dealing with forces and with what he called "Local Motions," was to form a rock-solid foundation for the emerging science of mechanics.

Galileo's work in mechanics began while he was yet in his teens. Though people have been watching weights swaying in the wind since the beginnings of mankind, the phenomenon had little significance until 19-year-old Galileo noted a surprising fact. While watching a church chandelier swinging in a breeze, he realized that the time it takes for one swing depends on the length of the supporting cable, not on the width of the swing, as would seem natural. This simple observation turned out to be the most important development in the early history of accurate timekeeping, and it led to the development of the pendulum clock.[20]

With a bit of luck, Galileo might have noted an additional fact about the pendulum. Not only does it swing back and forth in a plane, but also—if it continues to swing freely—it will change its direction of swing as the day progresses. What is happening is that Earth is rotating under the pendulum! This fact was not discovered until the nineteenth century and, ironically, it turned out to be the first solid physical proof of Earth's motion. Had Galileo noticed this phenomenon, he would have had the true demonstration that Bellarmine called for, and that Galileo so desperately sought in his work.

This demonstration was not to appear during Galileo's lifetime, however, and by the time of his trial, he was still vulnerable—so much so that he seemed an unlikely champion for any cause, even one so dear to his heart as freedom of scientific inquiry. Yet history has so placed him. Indeed, if any single event can be said to have created an enmity between science and religion, that trial and the resulting sentence was it.

Revisionist historians argue that the warfare between science and religion has been exaggerated; that it has really been a conflict between new science and any established authority[21]; that Galileo got what he deserved; that the trial was really a smoke screen intended to save Galileo from an even worse fate[22]; that there were other factors at work. Giorgio de Santillana suggested yet another approach to the conflict: While we can think of Galileo's ecclesiastical opponents as bigoted oppressors of science, he wrote, "It would possibly be more accurate to say that they were the first bewildered victims of the scientific age."[23]

All that may be. The fact remains that the Catholic Church is still suffering the effects of that fateful drama and is still trying to sweeten the bad taste many people sense in their mouths when the trial is mentioned; in the fall of 1980, Pope John Paul II ordered a new look at the evidence, the result of which, a dozen years later, was a very belated acquittal of Galileo. Further, the basic conflict—between established religion and modern science—is still being played out today.

To this day, if a speaker waves the flag of Galileo, we know immediately that the reference is to interference with freedom of scientific (or any) inquiry. Books are still being published and conferences are still being held on the causes, meaning, and outcome of the dispute.

Someday, if a group of astronomers at the Osservatorio Astrofisico in Arcetri have their way, such meetings might well be held in Galileo's villa, for the building still stands. Its name, "Il Gioiello" (The Jewel), is the same as it was 350 years ago. Unhappily, however, little else is; the building is sadly neglected. With special permission, one can walk through it, as I did some years back. It becomes an almost mystical experience to see the terrace where he looked up at the heavens, the small garden where he paced and pondered, and the various rooms that eventually became his universe: Toward the end of his life, he became totally blind, and his physical universe shrank down to what he could touch with his hands and fingers.

Among the visitors who managed to slip in to see him was Thomas Hobbes, who brought the news that the *Dialogue* had been translated into English. You'll meet Hobbes yourself in the next chapter.

Today, the villa stands closed, dark, untended. Members of the Osservatorio, which is connected with the University of Florence, would like to bring it back to life. Happily, Franco Pacini, the observatory's director, reports that restoration has begun[24]. He argues, however, that it would be wrong to make the villa a museum, which would be what he calls a "dead building." Rather, the hope is to resurrect it as a kind of living monument, perhaps as the Institute for Advanced Study in Florence, where scholars can come together and discuss new, and perhaps old, ideas in the world of science. Galileo would have liked that.

CHAPTER 2

Wallis versus Hobbes

Squaring the Circle

The seventeenth century in England was a time of religious and constitutional upheaval. Power struggles were endemic, complicated, and bloody, and the country began rolling toward revolution. Finally, in 1642, civil war broke out. While the major division was between supporters and opponents of the monarchy, the alliances actually involved a constantly shifting array of political, religious, economic, and even academic forces. In 1649, Charles I was beheaded by the Parliamentarians, and a short-lived Commonwealth was established.

Watching all this turmoil in utter dismay, and seeking desperately for some way to salve the wounds of his beloved country, was Thomas Hobbes—scholar, philosopher, and tutor to nobility. Though he was eventually to become famous and to attract an amazingly large amount of this hostility toward himself, he was born, in 1588, into the most modest of circumstances.

John Aubrey, a contemporary biographer of Hobbes, said that Hobbes's father "was one of the Clergie of Queen Elizabeth's time—a little Learning went a great way with him and many other ignorant Sir Johns in those days. [He also] disesteemed Learning . . . not knowing the Sweetnes of it."[1]

When Hobbes was seven years old, his father got into a fight with a fellow parson and was forced to flee from Malmsbury, Hobbes's birthplace, never to return. An uncle took over the boy's education and, apparently, did a good job of it. By age 14, Hobbes had shown a lively intelligence and was sent to Magdalen Hall (later named Hertford College) in Oxford. Like

Galileo, however, he felt uncomfortable with the standard scholastic curriculum (mainly the arts, including philosophy, and religion), and he read in other areas. His favorite subjects were geography and astronomy. He also began to develop an interest in optics.

"He did not much care for Logick," Aubrey continues, "yet he learned it, and thought himself a good Disputant. He tooke great delight there in Oxford to go to the Booke-binders shops, and lye gaping on Mappes."[2]

In 1608, the principal of his college recommended Hobbes for the position of tutor in the household of William Cavendish, who later became the Earl of Devonshire, then the Earl of Newcastle, and finally the Duke of Newcastle. The tutoring position was the first of several turning points in Hobbes's life, for it brought him into contact with a world of culture such as he had not known before. At the magnificent houses of the Cavendish family, he came to know the dramatist Ben Jonson, the poet Edmund Waller, and others of that intellectual caliber; he had at his fingertips a fine library, which, he maintained, was superior to the one at Oxford.

Sir Charles Cavendish, William's brother, was an accomplished mathematician, and William himself was a skilled amateur scientist, who maintained and worked in a well-equipped laboratory. In 1634, Hobbes sought a copy of Galileo's *Dialogue* for William, searching for it in London bookshops, though in vain. In a letter to William, he reported on his failure and his disappointment, commenting "I hear say it is called in, in Italy, as a book that will do more hurt to their Religion than all the books have done of Luther and Calvin, such [is the] opposition they think [exists] between their Religion and natural reason."[3]

In 1610, Hobbes and his pupil were sent on a grand tour of the European continent. Apparently, he learned more than his charge did, and on this tour, he decided to become a scholar. In the same year, Henry IV of France was assassinated, an event that seems to have made a deep impression on Hobbes.

On returning to England, Hobbes immersed himself in classical studies and, in 1628, he produced a translation of Thucydides's *The History of the Peloponnesian Wars*, which a modern reviewer has described as "majestic."[4] In the introductory section to this work, his political ideas were already beginning to take shape. "Thucydides," he wrote, "taught me how stupid democracy is and by how much one man is wiser than an assembly."[5]

This statement is jarring to our ears, but the idea must be taken in the context of his times. Hobbes, like other writers of this era, felt enriched and perhaps ennobled by classical history—including its ideals of heroism and its aristocratic politics. In addition, there were no examples of successful contemporary democracies that might have led him to think otherwise.

In Love with Geometry

On a second grand tour, undertaken in 1628, Hobbes had what can truly be called a transcendental intellectual experience. Again, Aubrey illuminates the life of Hobbes:

> He was 40 yeares old before he looked on Geometry; which happened accidentally. Being in a Gentleman's Library, Euclid's *Elements* lay open, and 'twas the **47** *El. libri* I. He read the Proposition. 'By G—,' sayd he (he would now and then sweare an emphaticall Oath by way of emphasis) 'this is impossible!' So he reads the Demonstration of it, which referred him back to such a Proposition; which proposition he read. That referred him back to another, which he also read. [And so on] until at last he was demonstratively convinced of that trueth. This made him in love with Geometry.[6]

Hobbes's contemporary, René Descartes, was also inspired by geometry. Descartes—who held back on publishing one of his works after hearing what had happened to Galileo—hoped that the whole world of physics might be reducible to geometrical quantities. Still today, Euclid continues to fascinate. Hideki Yukawa, winner of the 1949 Nobel Prize in physics for his meson particle theory, has written that by high school he was "captivated" by the beauty of Euclidian geometry. He, like Hobbes, turned to science only after this revelation.[7]

Hobbes's late conversion, however, was a likely cause of his later troubles. As Aubrey puts it: "Twas pitty that Mr. Hobbes had not begun the study of Mathematics sooner, els he would not have layn so open."[8] Like Yukawa and others who take to geometry, Hobbes was fascinated by the idea that a proposition, the truth of which is not obvious, can be deduced, by a chain of mathematically precise steps, from propositions that are accepted as true. This method, it seemed to him, could also be used both to build a complete philosophy and to demonstrate the logic of his own ideas. In fact, one of his works, the undated *Short Tract* (about 1630), is actually modeled on the propositional form of Euclid's classic work.

In other words, Hobbes wanted to be sure that his thoughts were not simply pumping up the fires he hoped to put out.[9] As he explained in a later book, *Human Nature:* "Those men who have written concerning the faculties, passions and manners of men, that is to say, of moral philosophy, and of policy, government and laws, whereof there be infinite volumes, have been so far from removing doubt and controversy in the questions they have handled, that they have very much multiplied the same."[10] He did not intend to make that same mistake. If his ideas were demonstrable (shades of Galileo), they could not be refuted.

A few years after Hobbes's conversion to belief in geometry, he embarked on a third tour of the European continent, bridging the years 1634 through 1637, during which he made contact with some of the leading scientific and mathematical minds of the day. In meetings with Mersenne, Gassendi, Roberval, and, especially, Galileo, he began to develop a deeper interest in the question of motion. After all, it was Galileo (whom he later hailed as the greatest scientist in history) who had shown that any aspect of the motion of a physical body can be expressed in mathematical terms. Basically, Hobbes felt that any natural occurrence is produced by some sort of motion, in fact that "it is by motion only that any mutation is made in any thing."[11]

Eventually, this idea became a cornerstone of his entire philosophical structure. He even regarded mental activities—including thinking and wanting—as actual, not metaphorical, motions of the mind. This belief made it possible, at least theoretically, to derive psychological insights from physical principles.

The idea is clearly a simple-minded approach to a far more complicated phenomenon. Its importance lay in its hubris, the suggestion that mental activities could be explained at all. The alternative, considering that nothing was known of brain cells, let alone of their operation, was reliance on the myriad superstitions and myths that had served as explanation for thousands of years.

Then, broadening the motion concept still further, Hobbes, like Galileo, began to think in terms of a grand design. Hobbes divided his design into three main sections. The first would treat "body," by which he meant material or substance, and its general properties; the second would have to do with humans and their special faculties and attributes; and the third would deal with civil government and the duties of subjects.

He had intended to write the entire work in proper order, but external events intervened. "It so happened," he later wrote, "that my country, some few years before the civil wars did rage, was boiling hot with questions concerning the rights of dominion and the obedience due from subjects . . . ; and was the cause which, all those other matters deferred, ripened and plucked from me this third part."[12] Thus was his masterpiece, *Leviathan,* born (1651). A brilliant, biting, forthright statement of political principle, it had the extraordinary effect of irritating just about everyone who read it—and many who didn't.

Leviathan

He starts his argument with a picture of humanity in the state of nature; this opening contains the words for which he is best known, words that have reverberated through history. In contrast with the romantic notion of natural man's condition, Hobbes describes it as "solitary, poore, nasty, brutish, and short."[13] It is a condition of competition and aggression, which is held in check, if at all, by fear of violent death. (One of his models for this primitive natural condition seems to have been life in America.) What a human in this state wants most from civil organization is protection. In return, the individual must give up some liberties. Hobbes's idea is still relevant today.

Hobbes compares the state, with its variety of parts, to some sort of large and frightening monster, Leviathan. Often thought of as a huge whale, Leviathan in mythology actually refers to any large serpentlike creature. What Hobbes is saying is that the state, like a giant monster, requires a single controlling intelligence if it is to operate effectively.

Monarchy, then, is to be preferred over any other form of government. In itself, this idea is hardly revolutionary. As an anticleric, however, Hobbes prefers a monarchical form of government for practical reasons, not because of any divine right, the usual foundation for the monarchy. He advises sovereigns not to permit any group or institution to form between them and their subjects, including, or perhaps especially, the Catholic Church. Also, whereas Descartes distinguished matter from spirit or soul, Hobbes the materialist argues that spirit does not exist, at least not in this world.[14] You can begin to understand why his very name was anathema to the church.

See where his line of reasoning takes him, however. Experimental philosophers, an emerging group who also had Galileo as their guiding light, were therefore equally dangerous, for they, just like the clerics, claimed an independent voice. Was Hobbes totally off base here? Isn't any discipline that is based on experiment a valid approach to understanding our universe? Not necessarily. In that inchoate world of early modern science, the practice of experimental philosophy was thought to be subservient to the higher level of the church's teachings. For example, Robert Boyle, a major figure in the new experimental philosophy, suggested that experimental philosophers were actually "priests of nature," and that their experiments should be carried out on Sundays, as part of their Sabbath worship.[15] This line of reasoning clearly does not mean that the early experiments were useless, but it certainly helps us understand Hobbes's suspicions.

Hobbes also had some pungent suggestions for reforming the education given to lawyers; in fact, he suggested reform of the universities in general. He felt strongly that the real purpose of these institutions was to develop scholastic arguments for advancing the cause of papal domination over civil power. He also believed that this underlying purpose was a powerful cause of unrest and dissension.

He felt too that the universities had fallen behind the times, and he argued forcefully for the inclusion of the science that had grown up outside of, and almost in spite of, the universities. Times were changing, however, and what had been true when he had attended university at the beginning of the seventeenth century was less true by midcentury. Thus, he was treading on very slippery ground.

Based on the workings of his beloved geometry, Hobbes also believed that he could reason his way to the truth, but using names instead of mathematical figures and numbers. Samuel I. Mintz, a Hobbes scholar at the City University of New York (now retired), suggests that in Hobbes's way of thinking, "True knowledge consists in reasoning correctly by names; that is to say, it is possible by a method of syllogistic reasoning closely analogous to computation in arithmetic to deduce correct conclusions from postulates and definitions."[16]

Hobbes's *nominalism* (belief that abstract concepts exist only as names, without referents) led him to *ethical relativism* (the belief that there are no absolute moral values or truths). "For True and False," he writes in *Leviathan,* "are attributes of Speech, not of Things. And where Speech is not, there is neither Truth nor Falsehood."[17] He also argued that it always makes sense to obey laws; in fact, he felt that without laws, it would not be possible to distinguish right from wrong. This idea hardly endeared him to his contemporaries; what then of divine guidance?

Who then was left on his side? Perhaps the Royalists? After all, he was a strong champion of absolute government, which one would expect to endear him to the hearts of the royal government. The problem was that he defended it not on the grounds of hereditary succession or divine right, but rather as the best way of protecting the ordinary citizen. He thereby managed to irritate the reigning powers as well.

Although *Leviathan* was not his first statement of his principles, it was the first to make a major splash. As a result, he was attacked from all sides. He was quickly branded an atheist, no light epithet in those days. He was the Monster of Malmsbury, the bugbear of the nation, the Apostle of Infidelity. He was the Insipid Venerator of a Material God, a panderer to bestiality, not to mention Anthropomorphist, Luciferian, Sadducean, and Jew.[18]

Nor did the attacks die down over time. In the early 1660s, a group of bishops in Parliament was calling for him to be burned as a heretic. This was no light matter, and Hobbes set about burning many of his writings, a fact bemoaned by editors of two major collections of his writings that have been published in the past few years. There were public burnings of his books, and the great fire of 1666 was laid at his door by some other members of Parliament, which called it a form of divine retribution for his doctrines.

What were Hobbes's own feelings about his work? Mintz includes in his book a quote sometimes attributed to John Bunyan, but which Hobbes is reported to have uttered: "I know there is a God; but oh! I wish there were not! For I am sure he would have no mercy on me."[19]

While Hobbes and his *Leviathan* quickly gained some admirers who also were tired of the endless strife in Britain, his opponents far outnumbered his supporters. Fortunately, many—perhaps most—were toothless. But some were not.

The Mighty Mathematician

Enter, on the other side of the ring, John Wallis—distinguished British mathematician, cryptographer, and cleric. Early in the Civil War he had deciphered some code letters for the Parliamentarians, which gives us a good idea of where his loyalties lay. Yet he somehow managed to stay on good terms with the monarchy when the Restoration (of Charles II) took place in 1660.

Wallis was 24 years younger than Hobbes. Though he studied a wide variety of subjects in his school years, including some mathematics, his major interest was divinity, and he was ordained by the Bishop of Winchester in 1640. In the next decade, he did some work in mathematics, particularly on the solution of algebraic equations.

In 1649, the position of Savilian Professor of Geometry, a highly prized post, fell open at Oxford when Royalist Peter Turner was dismissed, by order of Parliament. To the surprise of many, Wallis was appointed. Hence, mathematics, until then little more than a hobby for him, quickly became a serious occupation; within a few years, he was to emerge as one of the foremost mathematicians in Europe.

It is to Wallis that we owe the symbol for infinity (∞) and for less than or equal to ($\leq$). He also did work on the infinitely small, for which the symbol $1/\infty$ was useful. Newton, Lagrange, Huygens, and Pascal, among others, acknowledged the importance of Wallis's mathematical work. In fact J. F. Scott, an important biographer of Wallis, maintains, "When Newton modestly declared, 'If I have seen further it is by standing on ye shoulders of Giants,' he no doubt had the name of John Wallis well before his mind."[20] Wallis also did work in teaching the deaf to speak, in logic, in grammar, and in archival and theological areas.

Finally, Wallis helped to form (and was a solid member of) the Royal Society of London, an organization devoted to the advancement of science. While the society became and remains today a highly prestigious academy, it didn't start out that way. In *Letters on the English* (1733), Voltaire compared it with the Academy of Paris, and the Royal Society came out distinctly second best. As Voltaire put it, "Any man in England who declares himself a lover of the mathematics and natural philosophy,* and expresses an inclination to be a member of the Royal Society, is immediately elected into it."[21] *Except Hobbes.* For although this worthy scholar wanted very much to be a member (despite disclaimers), and although he certainly deserved to become one, he was successfully kept out by Wallis and several of Wallis's colleagues.

Wallis appears to have had a highly contentious disposition, as opposed to Hobbes who, although embattled on every side, was personally a far more pleasant person. Like Hobbes, Wallis engaged in a number of violent controversies. But he had a powerful mathematical pen, and the outcome of one such feud, with the widely respected French mathematician Pierre de Fermat during 1656–1657, helped cement Wallis's reputation in the field.

---

*The term *natural philosophy* was at the time what we might call observational or experimental science today.

Wallis's respectable reputation does not mean that he was always faithful to the truth, however. For instance, one section in a later book of his (*Treatise of Algebra,* published in 1685) has been called by science historian I. Bernard Cohen "one of the greatest distortions in the history of science." For Wallis, Cohen continues, "takes the point of view that all the major mathematics of the seventeenth century was developed by Englishmen, and that, for example, Descartes plagiarized from Harriot."[22]

It seems clear nevertheless that Wallis possessed wide-ranging interests and a powerful intellect. It seems clear, too, that he lay in waiting, like a spider in its web, for the hated Hobbes to blunder into his territory. Conveniently, Hobbes did so in 1655, at the age of 67, when he finally got back to work on his grand design. He published in Latin what was originally supposed to be the first book of his three-part work. There, buried in Chapter 20 of *De Corpore* (*Concerning Body*), was Hobbes's solution to a problem that had plagued geometers for more than 3,000 years—squaring the circle.

A Mathematical Challenge

Here is the problem: Draw a line with a straightedge. Place the point of a pair of compasses at one end of the line and, using the line as a radius, draw a circle. Then, using only the straightedge and compasses, measure out and construct, using a finite number of steps, a square that has the same area as the circle.

More academic/scholastic foolishness? Not at all. It's true that the problem probably was related in some way to the ancient Greek idea of the circle as a perfect figure. But it may also have had its origins in ancient Egypt, as an attempt to deal with a real-world situation. In fact, geometry itself seems to have begun there as a practical tool—a way of measuring and re-measuring plots of land, the borders of which kept disappearing under the annual floodwaters of the River Nile. The very word derives from the Greek *ge* (Earth) and *metrein* (measure). When a border has straight lines, the measurement problem is relatively easy; with curved borders, which are not unusual, it is much more difficult. Also, it would be handy to be able to reduce all such problems to one of measuring areas bounded by straight lines.

Within the Greek world of science and mathematics, however, any unsolved puzzle was a tantalizing challenge. Further, other, similar problems had been solved. For example, using straightforward geometric methods and the aforementioned straightedge and compasses, it was found possible to inscribe a triangle inside a circle, and then to double the number of sides as many times as desired. The same can be done with a circumscribed polygon. As the number of sides gets larger, the polygon gets to look more and more like a circle. In other words, the circle is the limit of these two series of polygons as the number of sides increases.

The method was known to Archimedes, who, using polygons of 96 sides, was able to show that pi is less than $3\frac{1}{7}$ and larger than $3^{10}\!/_{71}$.

Among others who have grappled with the circle-squaring (also called the quadrature) problem were the Greeks Anaxagoras, Hippias of Eli, Antiphon, Hippocrates of Chios, Euclid, and Ptolemy. It was tackled by the ancient Egyptians and Babylonians; by the Arabs and Hindus; and in the Christian world by Nicholas of Cusa, Regiomontanus, Simon Van Eyck, Longomontanus, John Porta, and Snell, as well as Christiaan Huygens, John Wallis, Isaac Newton, René Descartes, and probably Gottfried Leibniz.

Remember that in the mid-1600s, there was no calculus, now the very bedrock of much of our own science. Not only was geometrical thinking the order of the day, but the quadrature problem itself had become an object of widespread curiosity among the general populace—perhaps the only mathematical puzzle to do so. There were circle-squaring contests open to all, and the *Journal des Savants* (March 4, 1686) even reported that one "young lady positively refused a perfectly eligible suitor simply because he had been unable, within a given time, to produce any new idea about squaring the circle."[23]

With increasing interest, fired no doubt by the rise of the new science of Galileo and others, came a burgeoning of attempts to solve the puzzle—attempts by poorly trained mathematicians

who flopped about on the slippery mathematical ice, never re-
alizing how foolish most of their efforts were. The steady trickle
of solutions gushed into a flood so copious that in the eighteenth century, both the Royal Society and the French Academy
of Science issued decrees saying that no further attention
would be paid to such efforts.

It was into this morass that Hobbes fell. The heart of the
problem for Hobbes was that he himself had said he was basing
his philosophical ideas on his mathematical ones. If Wallis
could show that Hobbes's mathematical work was faulty, the
whole philosophical edifice might very well collapse.

Wallis later explained this tactic in a letter to the Dutch physicist and astronomer Christiaan Huygens (January 1, 1659):

> Our Leviathan is furiously attacking and destroying our universi-
> ties (and not only ours but all) and especially ministers and the
> clergy and all religion, as though the Christian world had no
> sound knowledge . . . , and as though men could not understand
> religion if they did not understand philosophy, nor philosophy
> unless they knew mathematics. Hence it has seemed necessary
> that some mathematician should show him, by the reverse
> process of reasoning, how little he understands the mathematics
> from which he takes his courage; nor should we be deterred
> from doing this by his arrogance which we know will vomit poi-
> sonous filth against us.[24]

Wallis, along with a colleague, Savilian Professor of Astronomy Seth Ward, resolved to take care of this Hobbes scoundrel
once and for all. Ward was to deal with the philosophical end of
De Corpore, and Wallis with the mathematical aspects. It took
Ward a year to respond, but Wallis moved in quickly for the kill.

Wallis later said that he was moved first to anger, then to
mirth, and finally to pity. But there was little pity in the slashing
refutation he issued just three months after *De Corpore* appeared. This was his *Elenchus Geometriae Hobbianae* (*elenchus* being the Socratic method of eliciting truth by cross-
examination). In this Latin pamphlet, Wallis tore into Hobbes's
definitions, as well as his methods; he probed and cut with
great skill—here using coarse mockery and there solemn ser-

monizing. He wrote of Hobbes's impetuousness and pride, of his danger to the church; Wallis even made fun of Hobbes's name, using a play on the words *hop* and *hobgoblin*.

Someone else might have wilted under this attack, but Hobbes followed the method of the brave defender—attack! He added a vituperative appendix to an English edition of his *De Corpore*—the sales of which, incidentally, profited from Wallis's attack. Hobbes titled the appendix, "Six Lessons to the Professors of the Mathematics, one of Geometry, the other of Astronomy." There was no doubt in anyone's mind that the two professors were Wallis and Ward. Starting right off in the dedication, he states that "from the seventh chapter of my book *De Corpore* to the thirteenth, I have rectified and explained the principles of the science [geometry]; i.e., I have done that business for which Dr. Wallis receives the wages."

A little further on, he refers not only to Wallis's *Elenchus* but also to two of Wallis's other books—"which two last I have here in two or three leaves wholly and clearly confuted. And I verily believe that since the beginning of the world, there has not been, nor ever shall be, so much absurdity written in geometry." In Lesson III of his appendix, he refers to Wallis's books as "mere ignorance and gibberish." In Lesson IV, he refers to one of them as "your scurvy book."[25] It is worth noting that in this "scurvy book" (*Arithmetica Infinitorum*, 1656), Wallis had taken a major stride toward what was later, in Newton's and Leibniz's hands, to become the calculus.

Lesson V is particularly revealing; he chides Wallis for writing of "a parallelogram whose altitude is infinitely little." "Is this the language of geometry?" Hobbes asks. One of Hobbes's problems, however, was that he was so taken with geometry as to be wholly blind to the new and growing capabilities of algebra. Therefore, he could, in good conscience, say of Wallis's ingenious methods regarding conic sections that it was "so covered over with the scab of symbols" that he had no patience to deal with it.[26]

He closes with, "So go your ways, you Uncivil Ecclesiastics, Inhuman Divines, Dedoctors of morality, Unasinous Colleagues, Egregious pair of Issachars, most wretched Vindices and Indices

Academiatrum."[27] It is worthwhile to decipher the last three terms, for, as Voltaire would later demonstrate even more vividly, a solid hit by wit could count strongly in a verbal duel. *Issachar,* a biblical reference, was used in the seventeenth century to describe a hireling who sacrifices principles for money. *Vindices*—that is, vindicators or protectors—is the plural of *Vindex,* the nickname Seth Ward adopted in his quarrels with Hobbes. Finally, *Indices Academiatrum* means traitors of the academy—Hobbes's pun on *Vindiciae Academiarum,* the title of a book by Ward defending Oxford and Cambridge from the charge made by Hobbes and others that the universities were centers of scholasticism and intellectual and scientific stagnation.[28]

Wallis, in reply, zeroes in on an error in Hobbes's use of a Greek word. The word Hobbes had used was *stigma* (in the original Greek, of course), which can mean a puncture or a brand mark. The word he should have used, says Wallis, is *stigmē,* which is a mathematical point—that is, a point without dimensions. To Hobbes, the words were interchangeable, mainly because he could not conceive of a mark without dimension. In the same way, he felt that a line must have width, and this was a basic difficulty in his solution of the circle-squaring problem.

Never daunted, he responds with "Marks of the Absurd Geometry, Rural Language, Scottish Church Politics and Barbarisms of John Wallis, &c." The mighty battle, in other words, has begun to degenerate into a wrangling about a wide variety of subjects, including some highly unimportant grammatical points, with both men feeling they have to display their erudition. Wallis thereupon replies in Latin with another play on words, "Hobbiani Puncti Dispunctio."

Hobbes held off on replying in 1657, mainly in order to complete his projected trilogy. Wallis, too, put the time to good use and came up with a comprehensive treatise on what might, in modern words, be called the foundations of calculus. It was published in that year and was titled, not unreasonably, *Mathesis Universalis.*

For a short time, all was quiet. Then, in 1660, Hobbes leaped back into the ring, issuing a detailed criticism of Wallis's works in the form of five Latin dialogues between two interlocutors, named A and B. Wallis later protested that A and B were none other than Thomas and Hobbes, and that their dialogue was nothing more than a method by which "Thomas commends Hobbes, and Hobbes commends Thomas, and both commend Thomas Hobbes as a third person, without being guilty of self-commendation."[29]

Hobbes jabbed back at Wallis in 1666, with the objective of beating down the pride of geometry professors. At this point, he confessed, he seemed to be battling "almost all geometers." He added wryly, "Either I alone am mad, or I alone am not mad; other alternative there is none, unless, perchance, some one may say that we are all mad together."[30]

By this time, the Royal Society was in full swing and was publishing its *Philosophical Transactions,* which, incidentally, continue to be published today. Wallis made good use of the transactions. In August 1666, he answered with "Animadversions of Dr. Wallis, upon Mr. Hobbes's late Book, *De Principiis et Ratiocinatione Geometrarum.* Written to a Friend," in which he turned Hobbes's musing about madness back on Hobbes. He argued that few will bother with a "confutation" of the book, for if what Hobbes said of himself is true, then the confutation "would either be needless, or to no purpose. . . . For, if he be the mad man, it is not to be hoped that he will be convinced by reason; or, if we all be so, we are in no capacity to attempt it."[31] Later, commenting on a point made by Hobbes, he wrote, "But why the crookedness of an arch should be called an angle of contact, I know no other reason, but because Mr. Hobbes loves to call that chalk which others call cheese."[32]

In 1669 Hobbes, more than 80 years old and apparently quite unaware of his true position in the matter, brought together his solutions to squaring the circle and to two other famous Greek geometry problems as well (cubing the sphere and doubling the volume of a cube geometrically). Again, as soon as they were published, they were attacked by Wallis. Again the verbal fists flew—in 1669, 1671, and 1672. The last was Wallis's final blast, but Hobbes was not finished. In 1678, at age 90, he came forth with his *Decameron Physiologicum,* a new set of 10 dialogues on physical questions. He could not forbear taking another swipe at Wallis, this time at some work Wallis had published on gravitation in his book *De Motu* (1669).

Hobbes died a year later. Born under the star of scholasticism, he had helped create a mechanical conception of nature. His was a deductive science, however, and when the members of the Royal Society moved on to the next step—that of experimental, inductive science—Hobbes could not keep up. So the great feud, which had gone on for almost a quarter of a century, died with him. Wallis, after 54 years as Savilian Professor of Geometry, died in 1703.

It will be interesting to compare this feud with the next two, namely those of Leibniz versus Newton and Voltaire versus Needham. In all three cases, a brilliant, wide-ranging philosopher and generalist was pitted against a narrower specialist. Such entanglements are far less likely to occur today, for science and mathematics have gotten so complex that few nonspecialists can find their way in them.

The outcomes of the feuds, as you will see, vary. In the case of Hobbes versus Wallis, the results were quite clear—to those who knew their mathematics. In spite of Hobbes's brave front, he always came out second best in the mathematical bouts with Wallis, yet Hobbes never seemed to recognize this.

Nor did his failures in mathematics seem to harm his reputation in other areas. The publication of *Leviathan* on the European continent provided him the fame he had always craved and created a widening circle of admirers, with whom he continued to correspond for the remaining years of his long life. He also received two glowing tributes from Leibniz in the early 1670s, in one of which Leibniz praised Hobbes for being the first philosopher to use the "correct method of argument and demonstration" in political philosophy.[33]

Hobbes would also have been happy to learn that his ideas had a powerful influence on the thinking of many leading scholars, including Spinoza, Leibniz, Diderot, Rousseau, Hume, and Locke. The years since World War II have seen a remarkable resurgence of interest in Hobbes's writings, reflecting recent attempts to find a way of dealing with the ever-increasing complexity of our lives, as well as the incredible power of modern military technology.

In fact, posterity has treated him far more kindly than did his contemporaries. He is sometimes called the first modern political philosopher. Hobbes scholar Michael Oakeshott has called *Leviathan* "the greatest, perhaps the sole, masterpiece of political philosophy written in the English language."[34] Thanks, too, to Hobbes's ideas on human behavior and how to deal with it, he is known in some circles as the father of scientific sociology.

Ironically, in the early pages of *Leviathan,* he had pointed out that "in Arithmetic, unpractised men must, and Professors themselves may often, erre, and cast up false." Then, later: "For who is so stupid as both to mistake in Geometry and also to persist in it, when another detects his error to him?"[35] So little can naïve tillers in the fields of science evaluate their own efforts.

In 1882, the German mathematician Ferdinand Lindemann was to show that the circle-squaring problem that so occupied Hobbes and Wallis was impossible to solve, as stated. Does this mean that the whole business was a waste? Mintz has called the controversy "futile,"[36] and Martin Gardner, writing in *Scientific American,* describes it as "profitless."[37]

Perhaps it wasn't as bad as all that. Repeated failures of geometers throughout the centuries, including Hobbes, forced mathematicians such as Wallis to seek elsewhere for solution—essentially to take a different route, through number and algebra, which eventually led to the next step, calculus.

Even Hobbes's philosophical ideas may have had some effect on the invention of the calculus. Carl B. Boyer, in his history of calculus, argues, "The excessive nominalism of Hobbes was to lead mathematicians away from a purely abstract view of the concepts of mathematics, such as Wallis had displayed, and to induce them to seek, during more than a century, for an intuitively, rather than a logically, satisfactory basis for the calculus." He adds: "It was largely on this account that both Newton and Leibniz sought to explain the new analysis in terms of . . . the generation of magnitude, rather than in terms of the logical conception of number only."[38]

CHAPTER 3

Newton versus Leibniz

A Clash of Titans

Calculus! The very word sends shivers down the spines of non-practitioners everywhere. What was, in Roman times, a small stone used as an aid in calculating has now grown to a rampart that must be breached by students in all the hard sciences, including of course mathematics, and in many of the social sciences as well.

Once having gained mastery over the subject, however, the student/scientist comes to recognize calculus as the most powerful instrument for scientific work that mathematics has ever produced. In fact, whereas *a* calculus has come to mean any process of computation or reasoning using symbols, *the* calculus refers to a specific method of analysis. While some inventions, such as the telescope and radar, sharpen the senses, others, such as logarithms and the calculus, increase the power of the mind. Even the computer, as used in science, cannot replace the calculus; it merely does the work faster.

Simultaneous Discovery

The calculus was discovered almost simultaneously by two men working independently of each other: Isaac Newton, an English scientist, and Gottfried Wilhelm Leibniz, a German philosopher. Not only did their feud have philosophical, religious, and diplomatic implications, but there were several other interesting outgrowths as well.

The feud may well, for example, have been a factor in the development of the *modern scientific paper,* meaning one that (a) is refereed or evaluated by the author's peers before it can be published, and (b) includes explicit, clear references to what has been accomplished previously as a way of clearly delineating what the author is actually contributing. This type of paper came into being around the mid–nineteenth century, after a long period of development, and its objective appears to have been less to share new discoveries with the rest of the scientific community than to provide a way of establishing the scientist's priority in his or her discovery.

Toward the end of the seventeenth century, however, scientific societies were still relatively undeveloped, and scientists often just circulated their papers—or letters or handbills— among a small group of peers. Both Newton and Leibniz did this with their first papers on the calculus, which turned out to be no help at all when solid evidence of priority was sought later on. In fact, it was not at all unusual for a new discovery to be announced in the form of an anagram; this reserved priority for the discoverer but could not be understood by anyone who did not already know the secret. Both Newton and Leibniz used the method.

That it did not work all that well as a way of establishing priority is shown by studies of sociologist Robert K. Merton, who found that in cases of simultaneous discovery in the seventeenth century, 92 percent ended in dispute. It is probably to the development of the scientific paper that we owe a decrease in contested priority over the centuries since the seventeenth. Merton has cited figures of 72 percent in the eighteenth century, 59 percent by the latter half of the nineteenth, and 33 percent by the first half of the twentieth.[1] Perhaps the pervasiveness of simultaneous discovery also became better recognized as time has gone by.

Even in the highly contentious seventeenth century, however, the Newton–Leibniz feud was something special, for it was truly a clash of titans. Both men were geniuses, even beyond the normal conception of genius. One of Newton's biographers, Richard S. Westfall, says that with other biographical subjects, he could measure himself against them—that in each case, there was at least a finite fraction. But, he says in his 874-page biography, *Never At Rest*, "The end result of my study of Newton has served to convince me that with him there is no measure."[2] Newton—who, by the way, was born in 1642, the year of Galileo's death—made major, fundamental discoveries in optics, mathematics, gravitation, mechanics, and celestial dynamics.

Leibniz, born four years later, is far less well known than Newton. Some would say this is in spite of, others because of, the feud. In either case, he was, in his way, both broader and deeper than Newton—and more modern. He has been called the last universal genius by historian Preserved Smith,[3] and the most comprehensive thinker since Aristotle by T. H. Huxley.[4] His interests included history, economics, theology, linguistics, biology, geology, law, diplomacy, and politics, as well as mathematics, celestial as well as terrestrial mechanics, and, equally important, philosophy. Frederick the Great of Prussia (Frederick II) called him "a whole academy in himself."[5] Yet he was not even an academic, as Newton was. He had been educated in law and supported himself by doing legal and diplomatic work for various members of the nobility in his native Germany.

Leibniz was also deeply interested in metaphysics, which was part of the reason that he and Newton could not come to terms. Yet it was this aspect of his philosophy that moved Leibniz, at least conceptually, beyond Newton and into realms that eventually culminated in what we know of today as modern physics. He did important work in symbolic logic, introduced an improvement in an early mechanical calculator, and played around with binary arithmetic, now the basis of our computers.

John Theodore Merz, one of Leibniz's biographers, described him as "of middle size and slim in figure, with brown hair, and small but dark and penetrating eyes. He used to walk with his head bent forward, which may have arisen from nearsightedness, or from his sedentary habits."[6]

Most of Newton's portraits were painted in his later years, when he had attained a position of prominence, and these, as was common, tended to idealize his appearance. What is clearly seen, however, is a great expanse of forehead, the stereotypical mark of the intellectual, and, in a few of the later portraits, the imperious look of the successful bureaucrat. The nose is long and thin, and the lower jaw somewhat receding.

His eyes have been described as "lively and piercing" by one contemporary, while another suggested that there was something "rather languid in his look and manner, which did not raise any great expectation in those who did not know him."[7] This dichotomy may reflect the feelings of the viewers; or perhaps the difference lies in whether Newton himself was deep in thought—which in this extraordinary man could reach an intensity matched by few other mortals. During his Cambridge days, his loss of contact with the outside world showed in his dress and habits, which were often careless, and in his neglect of food and even sleep when involved in a problem.

It is not surprising that so complex a man defies easy description; much also depends on the period of his life: Often pictured as dour and humorless in his earlier days,[8] he was also described as a delightful host by a group of French visitors at age 75.[9]

Foundations of the Calculus

Neither Newton nor Leibniz created their versions of the calculus out of thin air. By the mid-1600s, the basic components of the method were in existence, created by a variety of workers: In 1638, Fermat had discovered a way of finding maxima and minima in equations. Descartes's analytic geometry made it possible to have the equations of algebra stand in for the more cumbersome diagrams of geometry. And John Wallis's *Arithmetic* had established a link between the quadrature of curves (including the circle; see Chapter 2) and the drawing of tangents to them.

Note that drawing a tangent to a curve is a geometrical operation. (A tangent is a line that meets a curve at a single point without crossing it.) The angle that the tangent makes with the curve can then be physically measured. But, as was becoming clear in mathematical circles, the same result can be achieved algebraically, and more accurately, by creating a mathematical expression for that same angle.

In addition, however, a curve can be thought of as a path traced out by a moving point. Being able to deal with a moving point was important because the concept of motion occupied a central position in the philosophy of the time; it was thought not only by Hobbes but also by other philosophers to be the basis of all phenomena—mental as well as physical.

Hobbes, for example, had proposed the idea of *conatus,* or endeavor, a kind of impulse to both thought and action; it was the "beginning" of action. The concept involved not only instantaneous speed, a basic concept in the calculus itself, but also the pressure or motive force behind the movement.

Conatus, Hobbes suggested, is "motion made through the length of a point, and in an instant or point of time."[10] In other words, conatus is to motion what an instant is to time, what 1 is to infinity, what a point is to a line. Clearly, mathematics and philosophy were closely bound up in these questions, and a number of scholars, including Hobbes and Leibniz, were active in both fields.

Another problem of great interest had to do with measuring and calculating complex curves, areas, and volumes. The problem of determining the volume of wine casks, for example, had always been an important task, which had never been satisfactorily solved. Again, there was preliminary work to build on, including the so-called method of exhaustion, in which the area of a surface enclosed by a curve is found by inscribing polygons of an increasing number of sides. This, of course, is based on the same quadrature technique that Archimedes used in his work with pi (see Chapter 2). Similarly, a cone can be thought of as built up from a series of circles, each of slightly increasing (or decreasing) diameter from the next.

To the nonmathematician, it all seems highly esoteric. Voltaire, in his usual acerbic fashion, was later to describe the calculus as "the art of numbering and measuring exactly a thing whose existence cannot be conceived." Wallis, on the other hand, was able to move the technique along, using some brilliant work with infinite series. Newton studied Wallis's work in the winter of 1664–1665.[11]

In other words, specific instances of these types of problems had been handled both geometrically and algebraically by other mathematicians. So it is really no surprise that Newton and Leibniz developed the calculus independently at about the same time. A fascinating aspect of the discovery, however, is that they came at it from opposite ends; and in an odd way, the two approaches mirror the difference between *a* calculus and *the* calculus.

Leibniz, whose interests spanned many fields, was looking to create a unified system of knowledge. He was the holistic philosopher fighting a hopeless battle against specialization—a battle still being fought today. Toward this end, he had worked on a universal scientific language and became interested in what might be called a "calculus of reasoning." He yearned for a method that would facilitate his work with change and, in particular, with motion. This explains his interest in Hobbes's conatus. Leibniz, in other words, sought a general logical method— that is, *a* calculus. Perhaps it could even be used to unlock the secrets of human behavior.

For Newton, on the other hand, the calculus was more a way of dealing with physical problems, another mathematical technique added to the armamentarium of the physicist. Thus, he used it in working out many of the problems attacked in his most famous work, the *Mathematical Principles of Natural Philosophy,* commonly known as the *Principia* (1687). Then, apparently, he reworked the problems so that they could be presented in the traditional, mainly geometrical, fashion.

By mid-1665, he had developed the fundamental theorem of the calculus. By the fall of 1666, he had brought the method of "fluxions" (his term) to usable, if still somewhat clumsy, condition. He wrote a paper on the method and showed it to some colleagues, who urged him to publish it. Pulled on the one side by an understandable desire for recognition, he was yet beset by an almost pathological fear of criticism on the other. He refused to give permission for publication.

Thus, at age 23, and still a student, he had surpassed the leading mathematicians of Europe—and almost no one knew it. He then turned to other things. In 1669, thanks in part to his unpublished paper, he became Lucasian Professor of Mathematics at Cambridge University, a position that gave him the freedom to pursue his research interests.

A Toe in the Water

Hesitant from the first to feed his work to the scientific lions, Newton finally tried it in 1672, with a paper in the *Philosophical Transactions* of the Royal Society of London. Describing his first great discoveries in light and color, it was based on work he had done in the mid-1660s. At the close of the paper, he invited others to repeat his experiments, saying that he "should be very glad to be informed with what success," and that he would be glad to give further direction or to acknowledge his errors if he was found to have committed any. It was an invitation he came to regret.

Although the paper received a generally good reception and moved him into the limelight, it was also criticized. As a result, he found himself devoting precious time to answering often inane challenges to his ideas, a not uncommon result when one presented really new ideas. He soon began to complain about having sacrificed his peace. Among those objecting, however, were men of stature, including the Dutch physicist Christiaan Huygens and the British scientist Robert Hooke. He found their responses, especially those of the brilliant and contentious Hooke, extremely distasteful.

These early disputations seem to have had several curious effects on Newton. Though he continued to work on optics, he published no more papers on it; in fact, he held off on publishing his major work on the subject, *Opticks,* until after Hooke died—more than 30 years later.

Some of his communications were, however, read to the Royal Society, including several on subjects other than optics. These communications led to further problems with Hooke, which in turn may have goaded Newton to do some of his major research. One problem involved showing mathematically that masses attract as if they were concentrated at a point, and another had to do with showing that a planet orbiting the Sun under an *inverse square law of attraction*** will move not in a circle, but rather in an ellipse. Then, having proved his superiority to Hooke to his own satisfaction, he put the papers away somewhere and forgot about them for years.

He had tasted the world of public science and, finding it bitter, retreated into the island fortress of his Cambridge sinecure and his great mind. Largely because his early fears of publication were borne out, he consciously decided not to share his work freely with the world. He seemed to believe that his discoveries belonged to him and not to science—and certainly not to that vague repository, posterity.

Now for most discoverers, a strict determination of priority is clearly important, and Newton, in spite of some protestations to the contrary, felt the same way. Unlike many scientists, however, he felt that a scientist's priority derives from having done the work and not from publication of the discovery. Thus, when Leibniz's independent discovery of the powerful mathematical tool was published before Newton's, Newton discounted Leibniz's priority altogether. This difference of perspective was to be the cause of enormous conflict and heartache for both men in later years.

Embroilment

Newton began serious work on his *Principia* in 1684, which was just about the time that Leibniz began to publish information on his differential calculus. It was in the fall of that year that Leibniz's first paper appeared, in the German publication *Acta Eruditorum*. Newton's name was not even mentioned! Did Leibniz know about Newton's work? It seems likely that by that time he did.

Yet from Leibniz's point of view the omission was not unreasonable. Newton's mathematical reputation had been developing in England, but he still had absolutely nothing in print on any aspect of mathematics. Though Leibniz knew of Newton

**According to this law, two bodies attract each other with a force that varies inversely as the square of the distance between them. For instance, at a distance of 8 feet, the force is not 2 but 4 times weaker than at 4 feet.

through his travels as a diplomat, to most continental mathematicians the name would have been totally unfamiliar.

Let's try to imagine Newton's feelings, however. First, in spite of Newton's genius, his discoveries did not come easily, but rather were the result of constant, unremitting toil. As he put it, "I keep the subject constantly before me."[12] While it is true that he chose his lonely lifestyle, we cannot be sure that he felt no inner resentment about living such a solitary life. Then, to see someone else claiming full credit for the discovery—that had to hurt.

Second, although it is true that Newton had consciously decided against publication, he was surely aware of the importance of the discovery. Apparently he felt quite secure in the uniqueness of the work and was taken quite by surprise by Leibniz's publication. He was especially surprised that it was by Leibniz, for Leibniz had come to him for help eight years earlier! Newton had given it in the form of two letters, sent in 1676, in response to questions set by Leibniz.

Later, when the feud had escalated into true fury, these letters were to form the basis for one of Newton's major claims: He argued that he had shared some of his early work with Leibniz. In fact, however, although Newton had indeed written two letters to Leibniz in 1676, he had given away virtually nothing of the calculus itself. Nevertheless, Newton's suspicions of plagiarism arose at least partly because he found it difficult to believe that anyone could have advanced so fast from where he knew Leibniz to have been in 1676.

Newton had been scooped. Oddly, even this did not seem to bother him enough to force him into print. There was one small exception: He gave just a hint of the new method in the middle of his *Principia* (1687). The first real mention of his work on the calculus was only to appear in 1693, as part of a publication of Wallis's. Newton's own first publication of his calculus was a paper, "On Quadrature," which he had begun in 1691 and didn't get around to finishing. He had simply lost interest in it and put it aside. It finally appeared in 1704, as an appendix to his great book *Opticks*.

Note the date of Leibniz's publication: 1684. He, like Newton, was in no rush to publish. Though he didn't wait almost 40 years, as Newton did, he did hold off for 9 years. Each man, apparently, underestimated the other. Clearly, too, those were more leisurely days. Today's feverish rush to publish seems to have been missing then. Perhaps Leibniz felt some of Newton's fear of criticism.

There are two major parts to the calculus: differential and integral. Leibniz's second paper, on the integral calculus, appeared just two years after the first, perhaps prodded by publication of Newton's *Principia*. Here is how Leibniz described his own first publication in this second paper: "wherever dimensions and tangents occur that have to be found by computation, there can hardly be found a calculus more useful, shorter, and more universal than my differential calculus."[13] No false modesty here—and again no mention of Newton!

What it came down to, then, was that Newton did indeed discover the calculus first (1665–1666; Leibniz: 1673–1676), but Leibniz published it first (1684–1686; Newton: 1704–1736). By itself, this difference hardly sounds like the material for a feud of superhuman proportions. Perhaps if these two had been the only actors in the drama, they might have been able to reach some sort of compromise, for their relations at first were friendly enough. But there were other players in the wings.

Alliances

Neither Leibniz nor Newton had pupils to whom they passed on the calculus. The Swiss Bernoulli brothers, Jacques and Johann, however, took only a few days after Leibniz's 1684 paper to figure out the method, and they then proceeded both to use it and to pass it on to others. They quickly made contact with Leibniz and became his champions.

In fact, much of the rancor between the two discoverers came out of the pushing and prodding of the followers, Newton's as well as Leibniz's. Johann Bernoulli occupied a special place in the feud. On seeing something of Newton's calculus in Wallis's *Algebra*, he suggested that Newton had built it out of Leibniz's work. He referred to John Keill, a colleague of Newton's, as "Newton's ape." At other times, he called Keill "Newton's toady" and "a hired pen."

While he might do this in writing, however, he never used Keill's name but rather referred to him as a "a certain individual of the Scottish race."[14] In fact, Johann Bernoulli's fighting style generally came down to, "Let's you and him fight," for he constantly prodded Leibniz into combat but did everything he could to stay out of the front lines and to remain anonymous. Later on, he even tried to establish friendly relations with Newton himself.

Newton, too, had his partisans, whom Leibniz came to call Newton's *enfants perdus* or reconnaisance patrol, but none were of the intellectual level of the Bernoullis. Wallis, for instance, had certainly been a first-rate mathematician, but he was getting old and had passed his prime; his creative energies were also taken up in his squabble with Hobbes.

Nevertheless, Wallis was deeply concerned that the Germans, for whom he held no love, would pull ahead of the English in mathematics and science. He urgently prodded Newton to publish his calculus, saying in 1695 that Newton's discoveries already pass elsewhere "with great applause by the name of Leibniz's Calculus Differentialis. . . . You are not so kind to your Reputation (& that of the Nation) as you might be when you let things of worth ly by you so long, till others carry away the Reputation that is due to you."[15]

At this late date, however, Wallis's fears were already coming to pass. In fact, Leibniz's followers were to prove better at applying his method than he was himself; as a result, this group of European continental mathematicians dominated the mathematical scene for the next generation. Included in the group, in addition to the Bernoullis, were such respected names as L'Hôpital, Malabranche, and Varignon (a French mathematician who was courted by both sides, and who later moved over to Newton's corner).

Also, it was to turn out that Newton's notation was somewhat less convenient to use than was Leibniz's; in fact, we still use Leibniz's form today (e.g., dy/dx). Nonetheless, the British mathematicians, dazzled by the glory of their master, were blind to this difference in ease of use, so they continued to use Newton's rather more cumbersome dot notation. This excessive deference to their master was to hold back British mathematics for a century.

In the meantime, it seemed that just about the only way for Newton to gain back some lost ground was to show either that Leibniz had, to put it bluntly, plagiarized, or that Leibniz's formulation was inferior to his. In fact, John Wallis, David Gregory, John Collins, and other followers of Newton were convinced that Leibniz was indeed guilty of plagiarism. The most likely occasion for the theft was in October 1676, when Leibniz visited London, and Collins showed him some of Newton's unpublished papers. Modern scholars have been able to examine Leibniz's notes on the meeting and are convinced that he did not actually build his discovery on these items. Nevertheless, Newton's followers persisted in believing that on that occasion, their hero was robbed. Leibniz was later to complain of Wallis's "amusing affectation of attributing everything to his own nation."[16]

On the other hand, when Leibniz published his calculus in 1684, he made no mention of Newton's two 1676 letters, nor of the fact that he had seen some of Newton's unpublished papers, courtesy of Collins. He posed, in other words, as the sole inventor of the method, and he held that pose for some 15 years. Though it is probably a coincidence, there is also something suggestive in the fact that Leibniz's first paper was published just one year after Collins's death in November of 1683. There has even been a suggestion of a conspiracy. In 1920, Arthur S. Hathaway put forth the idea that Collins was a German agent, acting on behalf of Leibniz and the honor of the German nation. Though the idea sounds more like science fiction than science history, it was published in no less a journal than *Science*.[17] In any case, Newton finally learned that Leibniz had seen his papers. This revelation convinced him that Leibniz knew he had the calculus and still gave him no credit; this he could not forgive.

Still, Newton was clearly remiss in keeping his discovery close to his chest. When we consider the question of who actually brought the calculus to life and made it available to others, the credit must clearly go to Leibniz. In later years, when the controversy was in full swing, Leibniz was to write, "The inventor [meaning himself] and the very learned men who employed his invention, have published beautiful things which they have produced with it; whereas the followers of Mr. Newton have not effected anything in particular, having hardly done more than copy the others, or wherever they wanted to pursue the matter have tumbled into false conclusions. . . . Hence it can be seen that what Mr. Newton has found is to be attributed more to his own genius than to the advantage of the invention."[18] Aside from the rather exaggerated statement about Newton's followers copying the others, the comment is not far off the mark.

The Royal Society

Nevertheless, even though Newton no longer pursued research in mathematical physics, his earlier discoveries and publications had carried him to the forefront of the scientific world. He had, in addition, become Master of the Mint in 1699—an important position he carried out with expertise, conviction, and zeal. The conversion from a solitary, haunted scholar to a self-assured, forceful, and somewhat paunchy bureaucrat had been completed. He even served for a short time in Parliament, though without distinction.

Along with the change came an interest in the Royal Society as a place for scientific exchange. Hooke, however, had been its president and mainstay for many years. Newton was interested enough to be on the Royal Society's council but did not play an active part until Hooke's death in 1703, after which he had an open field.

He became president in that year and was soon proposed as "perpetual dictator" by a number of members.[19] Newton scholar Frank E. Manuel has called Newton "the first of a new type in European history—the great administrator of science."[20] It is a role that casts light on the position of those who "run" science, as opposed to those who "do" it.

Newton made the society his own. He packed its governing body with his friends and colleagues and began to use its name and those of many of its members as the shields and swords of his further controversies. Apparently, he saw and helped write many of the manuscripts put forth by his disciples, including an anti-Leibniz preface to the second edition of his own *Principia*.

On the other hand, the society had been in bad shape when he took over; officials attended irregularly, and members' dues were commonly in arrears. He converted it, almost single-handedly, from what had come to be a joke in some circles to a respected institution, one to which many non-English practitioners—aristocrats as well as scientists—sought admission.

Newton even managed to move the society into a new building. During the move, a portrait of Hooke—painted while he was president of the society and hanging on the walls of the old building—simply disappeared. Thus, there is not a single known portrait of Hooke in existence today.[21]

The eighteenth century had opened with a bang, or perhaps a continuous rattle would be more appropriate. Both Newton and Leibniz, preeminent in their respective domains, were passing insulting remarks about each other and about each other's followers and were encouraging their followers to do the same in scholarly journals. The main journals of the day were the *Philosophical Transactions*, Newton's domain, and the *Acta Eruditorum* of Leipzig, over which Leibniz had some influence.

Leibniz, however, was also a member of the Royal Society, which led to an extraordinary situation. Among the insinuations being passed around was John Keill's strong implication of plagiarism by Leibniz, published in the *Philosophical Transactions* in 1708. Keill, a competent though aggressive mathematician, had written in his paper that Newton's priority existed "beyond any shadow of doubt."[22] This assertion could only be taken as an implied accusation of plagiarism. Keill was, of course, a member of the Royal Society and one of Newton's *enfants perdus*.

Leibniz hesitated for a while then decided to challenge Keill. In 1711 and 1712, he sent two virulent letters of protest to Hans Sloane, secretary of the society, complaining of the insult.[23] Leibniz's letters really pushed Newton over the edge. From that time on, he devoted a large part of his time and effort to a continuous detailing of his case against Leibniz. From then on, it was war, and Newton was relentless in his campaign.

The society, in response to Leibniz's challenge to Keill, convened a panel to look into the matter. Newton insisted that the panel was impartial and later wrote, "The Committee was numerous and skilful and composed of gentlemen of several Nations."[24] In truth, it consisted, with one exception, entirely of Newton's partisans. In fact, its makeup was so transparent that the names of the committee members were not even published when the original report was made.

The report, a long and detailed review of the situation, was ready in the surprisingly short time of 50 days and contained information that could only have come from Newton himself. In fact, a draft of the report exists in Newton's own writing. Not surprisingly, it was highly favorable to Newton's cause and put Leibniz in a very awkward position.

In an account of the whole situation, again written anonymously by Newton and published by the society, he turned the tables completely. In answer to the charge that the society had given judgment against Leibniz without hearing both parties, he argued that this charge was mistaken, that the society had not yet given judgment in the matter. In fact, he wrote in all innocence, it was Leibniz who was at fault for wanting the society to condemn Keill without hearing both parties. In so doing, the anonymous Newton threatened, "he has transgressed one of their Statutes which makes it Expulsion to defame them."[25] Newton was later reported by a colleague to have said "pleasantly" that "He had broke Leibniz's heart with his Reply to him."[26]

Other Factors

To help us understand how these two great men could have gotten themselves involved in this less-than-lofty imbroglio, it is necessary to look a little more carefully into their personal lives and at their philosophical and religious views.

Leibniz's Worldview

Newton's name has become a household word, so, in spite of his extraordinary brilliance, it is possible to think of him as a human being. Leibniz, however, remains a curiosity, a disembodied star that hangs like many others in the heaven of philosophy. He is different from the rest of us, distinct, as the celestial laws were distinct from the terrestrial ones in Galileo's day.

Yet, if we could but enter his world, we would see, indeed feel, a life of pain and disappointment, of unresolved controversy. John Theodore Merz wrote, "In spite of the many controversies in which Leibniz was entangled, it does not appear as if he had been really victorious in any one of them. Many of his antagonists did not think it necessary to reply to his remarks, others broke off the lengthened discussion, at times the death of the controversialist brought the undecided argument to an end. Thus not only the personal quarrel about the infinitesimal calculus remained without distinct issue, but Leibniz's arguments with Arnaud, Bossuet, Locke, Clarke, Bayle, and many others, led to no decided victory."[27]

He occupied no position of influence or power, as Newton did. He suffered severe rheumatic pain, and from 1676 until his death he served as librarian, judge, and minister for the house of Brunswick in Hanover, Germany. To these, he had to add the job of historian and genealogist when he was given the task of compiling a dynastic history in support of the Brunswick family's imperial claims to the British crown—hardly a task worthy of an intellect such as his. He described himself as living in a state of distraction among historical, philosophical, and scientific research.

Somehow, his many schemes never seemed to come to fruition. He had worked on plans for unifying the Catholic and Protestant Churches, with obvious lack of success. In 1672, he journeyed to Paris to present Louis XIV with a political scheme that involved a French attack on Egypt as a way of weakening the Ottoman Empire and deflecting French aggression away from Germany. Nothing came of it.

Further, Germany, which had earlier flourished in the arts and sciences, now exhibited a breakdown in manufacturing and trade, in military power, in government, and in the arts and sciences. Using the examples of England, France, and Italy, which had active academies and learned societies, and believing that these could do much to advance the cause of learning in his country, he worked on an idea for a series of learned societies.

In 1697, a plan for an academy in Berlin was sanctioned by the Elector of Brandenburg, later the first King of Prussia, and Leibniz was to head it. Unfortunately, a series of wars broke out and drained away whatever energies and money might have gone into such an academy. A similar plan at Dresden similarly foundered, again for political reasons—and again in Vienna! Of the many projects Leibniz propounded, only the Academy of Sciences in Berlin came into existence during his lifetime, though only barely, and he served as its president.

Newton's Worldview

Newton, too, had his troubles. He never knew his father, who died before he was born, and he could be said to have lost his mother to his stepfather. He, like Leibniz, never married and had no close family. The years of loneliness and of operating at fever pitch apparently took their toll, for he had a serious mental breakdown in 1693. It lasted for about a year, during which he accused friends of plotting against him, and he had great difficulty sleeping.

He also imagined conversations that never took place. In a letter to Samuel Pepys (a diarist of that era), dated September 13, 1693, he wrote, "I am extremely troubled at the embroilment I am in, and have neither ate nor slept well this twelve-month, nor have my former consistency of mind. I . . . am now sensible that I must withdraw from your acquaintance, and see neither you nor the rest of my friends any more."[28] A sad letter indeed. A few days later, he wrote to John Locke, apologizing for "Being of the opinion that you endeavoured to embroil me with woemen," and for calling Locke "a Hobbist."[29]

That the breakdown took place is not in question. The cause, however, has been a matter of debate for years. Various possibilities have been suggested: the death of his mother (though that was years earlier, in 1679); his failure to obtain certain posts he was interested in; and a serious loss of manuscripts in a fire, though this loss has not been authenticated.

Recent chemical tests of Newton's hair suggest another intriguing possibility. Due to an intense involvement with alchemy from about 1669 onward, Newton had long been exposed to the vapors of a wide variety of poisonous chemicals, and in fact he was known to have used his sense of taste as one of his tests for the chemicals. A high concentration of mercury in his hair suggests that he may have given himself a severe case of mercury poisoning, for prior to the breakdown he had been involved in a series of alchemical experiments on which he labored far into the night, as he often did when involved in a project. At times, he would simply fall asleep in the presence of a pot of boiling mercury.[30]

In later years, he was too busy to spend that kind of time with his retorts, which may have saved his life. He maintained an extraordinary ability to solve physical and mathematical problems when these were presented as challenges by his adversaries, which was common in those times. Nonetheless, the days of his earlier creative brilliance—during which he discovered the binomial theorem, the calculus, the constitution of white light, and the theory of gravitation—were no more. Further, the years 1711–1714 were marked by a struggle for control at the British mint, and by conflicts with the first Astronomer Royal of England, John Flamsteed, on other matters at the Royal Society.

His religious convictions may have added to his internal tensions. He followed Arianism, a Christian sect that denied the predominant belief in the Trinity—and which was heartily loathed by most of his British compatriots. William Whiston, a colleague and fellow Arian, revealed his own convictions after Newton's death and was promptly thrown out of his job as successor to Newton at Cambridge.

So we have Leibniz and Newton facing off, one blocked at almost every turn, particularly in his later years, the other imperious—nouveau noblesse, if you will—and perhaps just a little unbalanced. The results of the duel were to reverberate through the world of natural philosophy.

Philosophy and Religion

At this point in our discussion, the feud begins to take on overtones reminiscent of the Wallis-Hobbes dispute, for a large part of the animosity between the two men derived from differences in their basic ideas, both philosophical and religious. Though Newton had been a card-carrying alchemist since 1669, he was able to keep that part of his life quite separate from his mathematical physics. To the world of science, he was the strict empiricist. In the *Principia,* Newton had provided a single mathematical explanation for phenomena as diverse as the movement of the planets, the motions of the tides, the swing of a pendulum, and the fall of an apple. This astounding feat finally pulled together terrestrial and celestial physics, a job that Galileo would have liked to do but couldn't.

The single driving force in all these activities was gravity. Newton cautiously and sensibly refrained from trying to explain just what this gravity was—a full explanation of which is still forthcoming. Implicit in its use, however, was acceptance of the "action-at-a-distance" concept.***

Action at a distance came to be used in studies on electricity and magnetism as well, for all of these phenomena operate under the inverse square law (see footnote **, earlier), upon which Newton based his work. In other words, the concept lent itself to mathematical treatment, and this is what made it acceptable in the world of science.

The continental mathematicians and philosophers, however, found the idea of action (in this case attraction) at a distance just too much to take. Leibniz in fact felt a need for some sort of subtle matter in space to explain the planets' motion. See now how the science and metaphysics of the time intertwined. Leibniz was supposedly the metaphysician, yet to him and his confreres, action at a distance smacked too much of the occult; it had a "fantastical scholastic quality." It was, they felt, a long step backward. At the same time, they attacked Newton for not trying to explain what gravity is or how it works.

Summarizing the difference of opinion between himself and Leibniz, Newton wrote, "It must be allowed that these two Gentlemen differ very much in Philosophy. The one [himself]

***Action at a distance refers to two objects separated in space, with each having some sort of effect on the other, without any apparent physical connection between them.

proceeds upon the Evidence arising from Experiments and Phaenomena, and stops where such Evidence is wanting; the other is taken up with Hypotheses, and propounds them. . . . The one for want of Experiments to decide the Question, doth not affirm whether the Cause of Gravity be Mechanical or not Mechanical: the other that it is a perpetual Miracle if it be not Mechanical."[31]

Closely allied were other, even more irritating, differences. Both Leibniz and Newton were deeply religious, but their feelings about the part played by God in our universe diverged widely. If the universe does indeed operate on strictly mechanical principles, then, said Newton, the universe could be thought of as a clock, one that God wound up at the beginning of creation.

He feared, however, that if the clock ran on forever without God's help, the need for God becomes superfluous. If God had simply wound up the machine and let it go, what could prayer accomplish? He felt, for example, that certain unexplained irregularities in the motions of the planets might add up and finally throw the whole solar system out of kilter. In such case, he was confident, God would step in and set things back in order.

Leibniz, on the other hand, derided the idea that God was some sort of astronomical maintenance man. He felt that a universe that ran down like a clock, needing periodic rewinding, was totally unacceptable, that such a conception of the universe derogated God's perfection.

This belief in the perfection of God and the universe formed an important part of Leibniz's philosophy. He believed that God had carefully chosen among an infinity of possible worlds the one He felt was most suitable. So we may not have a perfect world, but, all things considered, we have the best of all possible worlds. As we'll see in the next chapter, the idea was to be viciously satirized by Voltaire.

The two feuders' views on Hobbes were in opposition as well: Leibniz found Hobbes's philosophical views useful and palatable; Newton abhorred them. (Recall his slanderous attack on Locke as "a Hobbist.") For the modern reader, however, the most interesting aspect of the whole dispute is probably the difference between Newton's and Leibniz's views of space and time. To Newton, space and time were absolute and real entities. They existed independently of the human mind. This certainty provided a solid foundation on which science built what has now come to be called "classical physics"; and until the advent of twentieth-century relativity theory, physics existed in a Newtonian universe.

Leibniz had a totally different concept of space and time. He felt that if they were absolute and real, they would be independent of God and in fact would be setting limitations on God's capabilities; that is, God would not be able to exert any control over them. Again, it looked as though Newton the empiricist opposed Leibniz the metaphysician.

Yet, every so often, Newton's nonempirical views crept out into the open. In an early edition of his *Opticks*, for instance, Newton had suggested that space was some sort of "sensorium" of God. He quickly changed his mind and tried to call back the copies that had been released, to replace that statement with new text. One of the original copies fell into Leibniz's hands, however, and he tore the idea to shreds. Does God need sensory organs in order to perceive?

Leibniz also pointed to what he called the anti-Christian influence of the *Principia*. Newton found this harder to take than the calculus feud, partly perhaps because it was not Leibniz's idea alone. A Pietist named Franke said he could not make good Christians out of geometry students, and someone called Wesley ended his study of mathematics because of his own fears that it would make him an atheist.[32]

Leibniz's ideas on space and time had some interesting implications. For him, space and time were orders, or relations. Space was the "order of coexistences," time the "order of successions." If all bodies in the universe doubled in size overnight, Leibniz asked, would we notice anything different the next morning? He said we would not; the size of our bodies having doubled, he argued, there would be no way we could discern a change—and this was in the early 1700s.

"So profound was he," wrote historian Preserved Smith, "that his doctrines did not come into their own until the rise of Relativity."[33] It took two centuries for physics to catch up with these relativistic views of Leibniz. Einstein and others in the line of relativistic succession found Leibniz's ideas useful.

Leibniz even argued against Newton's hard, "massy" particles as the constituents of matter. He replaced these particles with a set of *monads*—entities without extension, parts, or configuration, yet which possess in infinitely varying degrees the power of perception. To the hardheaded realist, his monads sound impossibly metaphysical; Newton derisively called them "conspiring motions." Yet even "conspiring motions" come closer to the quantum mechanical conception of atoms than do the "massy" particles of Newton.

Science historians sometimes ask whether it would have been better had the two men been able to work together. In a certain sense, however, they did, for their ideas played off one another's. Though they went off in many opposite directions, the calculus dispute bound them together into a package that was eventually to spring forth into what we have come to know as modern physics.

A Battle to the Finish

Still, there is no doubt that in terms of the feud, Newton got the better of Leibniz. Thanks at least in part to the job Newton and his cronies did on him, his star, which had shone so brightly, began to dim, and then, for a time, went out altogether. Toward the end, the disputants' situations could hardly have been more different. Newton was respected and admired. He had been knighted, the first person to receive that honor for contributions to science. When he died in 1727, he was given a state funeral, and he still lies buried in a prominent position in the nave of Westminster Abbey.

For Leibniz, nothing seemed to go right. Ironically, in 1714, when the Elector of Hanover, his employer, became George I of England, Leibniz even lost favor in his own court. Again, this was probably a result of his duel with Newton. The calculus feud had become a factor in the diplomatic maneuvering between Great Britain and Hanover, and Leibniz was clearly on the losing side of the feud. Who wants to be associated with a loser? He had even tried to get the Roman Curia to release Galileo's *Dialogue* from the index, also without success.

When Leibniz died in Hanover in 1716, unfulfilled in his many schemes, and with hardly a friend in the court where he had labored for almost four decades, his funeral was attended by no one other than his former secretary. A friend noted in his memoirs that Leibniz "was buried more like a robber than what he really was, the ornament of his country."[34]

He died, said Merz, "in the gloomiest period of his country's history, and in a world full of deception, ruin and wretchedness."[35] Yet, somehow, his work maintained a powerful optimism that showed up in several ways, not the least of which was his "best of all possible worlds" idea. In fact, he was among the first to break from the feeling that civilization was suffering a continuous and inevitable decline from an ancient golden age. The eighteenth-century philosopher Diderot called Leibniz the father of optimism.[36]

Considering all the personal disappointments he suffered, this optimism is all the more remarkable. Considering, too, that Newton's name has been honored by being used as a unit of force, which seems appropriate, I propose that it is time to similarly honor Leibniz as well. I would make the *Leibniz* a unit of optimism. Perhaps some latter-day combination of Leibniz and Newton will arise who will show us how to quantify this unit.

CHAPTER 4

Voltaire versus Needham

The Generation Controversy

"Gentlemen, between two servants of Humanity, who appeared eighteen hundred years apart, there is a mysterious relation," said Victor Hugo. "Let us say it with a sentiment of profound respect: Jesus wept; Voltaire smiled."[1] Voltaire's smile had a thousand faces, however—and as many uses. He used it to fight injustice, intolerance, and absolutist power, in both church and state. He used it for personal vendettas as well, and no one else could apply it with such devastating effect. "Ridicule," he maintained, "overcomes almost anything. It is the most powerful of weapons. To laugh while taking vengeance is a great pleasure."[2]

During his long lifetime, which spanned the first three quarters of the eighteenth century, he produced a flood of poems, letters, plays, histories, and political pamphlets and stories. Some he signed, some he didn't, in order to avoid attribution. In his day, in France, many statements could be made on paper that would spell torture or death if authorship could be proved. He had in fact spent time in the Bastille, as well as many years in exile from his beloved Paris.

Voltaire was the master of nuance. After a run-in with Frederick II, King of Prussia (often called Frederick the Great, though probably not by Voltaire), he wrote in a letter to his niece that he was compiling a little "Dictionary for the Use of Kings," and he gave a few samples. "My dear friend" means "you are absolutely nothing to me." By "I will make you happy," understand "I will bear you as long as I have need of you." "Sup with me to-night" means "I shall make game of you this evening."[3]

One never knew where Voltaire would strike next. One of his major targets was P. L. M. (Pierre Louis Moreau) de Mauper-tuis, a leading figure in eighteenth-century European science, as well as president of the Berlin Academy of Sciences—the very academy that Leibniz had hoped to head half a century earlier. Maupertuis had been among the first major figures to recognize Newton's worth and had been sent on a difficult but ultimately successful mission to check out one of Newton's theoretical predictions.

In the early 1750s, Maupertuis was involved in a dispute with a mathematician named Koenig. Using the same technique perfected 40 years earlier by Newton, he convened an academy panel that was intended to railroad Koenig. In his defense, Leibniz had had no avenging angel, but Koenig had one: Voltaire, who had a variety of reasons for disliking Maupertuis, leaped to Koenig's defense. His technique was simple but brilliant—make a fool of Maupertuis and thereby destroy his credibility.

In 1752, Maupertuis had written up some ideas in a series of published letters. A few were sensible and useful, but several were outlandish. He suggested, for example, blowing up the pyramids of Egypt to find out what is inside them; founding a city in which only Latin should be spoken; digging a pit to the center of Earth, to see what is down there; and vivisecting criminals who had been condemned to death, believing that in dissecting the brain, the mechanism of the passions could be elucidated.[4]

Voltaire pounced on these ideas and, through them, on Maupertuis. The basic weapon was an essay titled "A Dissertation by Dr. Akakia, Physician to the Pope."[5] In this delicious satire, Akakia expounds on the actions of a conceited young student who writes "Lettres" and tries to pass himself off as the respected president of an important academy. Maupertuis's name is never mentioned, but there is no question about whom the author (not Voltaire, of course) has in mind. Part of the work consists of an inquisitorial examination reminiscent of the one aimed at Galileo and his *Dialogue*, but in this case, the examination is of the "Young Author's" letters.

A sample: "We pass over several things that would weary the reader's patience, and are unworthy of the inquisitor's notice; but we believe he will be greatly surprised to hear that this young student is positively anxious for dissecting the brains of giants . . . , and of hairy men with tails, the better to discover the nature of the human mind; that he proposes to modify the soul with opium and dreams; and that he undertakes to produce . . . fishes with grains of dough." Then, after describing more such foolishness, the examiners add, "To conclude, we entreat Doctor Akakia to prescribe to him some cooling medicine; and we exhort the author to apply himself to his studies in some university, and to be more modest for the future."

When Maupertuis sent a note to Voltaire threatening vengeance, Voltaire printed a further adventure of Akakia and put Maupertuis's letter at the beginning. With his extraordinary facility for finding the point, sharpening it to razor sharpness, and driving it home, he quickly made Maupertuis the laughingstock of all Europe. It was all too much for Maupertuis, and he died a few years later, broken in spirit, in health, and in mind.

Though not a scientist himself, Voltaire had developed a keen interest in both the physical and the biological sciences. In fact, it was largely through Voltaire's *Elements of Newton's Philosophy*, published in 1738, that Newton became widely known on the European continent. (Ironically, it had been Maupertuis who first introduced Voltaire to Newton's work.) Voltaire was, in other words, one of the earliest and best of our science writers—meaning someone who can convert complex scientific material into readable prose.

In writing his book on Newton, Voltaire had some very unusual help: In the years 1733–1749, he had as a lover and colleague the remarkable Gabrielle Emilie le Tonnelier de Breteuil, Marquise du Châtelet-Lomond, who managed to combine in one package riches, charm, and brains. In fact, working with her tutor Maupertuis, she had developed a better understanding of Newton's work than had Voltaire. Worse, for a while, she was strongly supportive of Leibniz, whom Voltaire disliked. Still worse, urged on by Maupertuis, she had produced a work on the German philosopher. Perhaps worst of all, Emilie, though a married woman and Voltaire's lover, had developed a passion

for Maupertuis—which, fortunately for Voltaire, was not recip-
rocated and passed after a while.

In 1759, seven years after Voltaire dispensed with Maupertuis
via his *Akakia,* and well after Emilie's death in 1749, he turned
his sights on Leibniz. In the process, he produced his most famous single work: *Candide,* a savage satire on eighteenth-
century life and thought; on religious fanaticism, war, and the
injustices of class distinction; and, finally, on the philosophy of
Leibniz. While the hero of the story is Candide, his mentor is
Dr. Pangloss, a disciple of Leibniz's. In the face of an extraordinary series of seriocomic adventures, Pangloss maintains, as
did Leibniz, that "All is for the best in this best of all possible
worlds." The book was one of Voltaire's greatest successes, and
countless millions of copies of the story are still in print, in
many editions, the world over.

Voltaire had two major objections to Leibniz's work. First, he
felt that Leibniz's "best-of-all-possible-worlds" position was totally misleading, that Leibniz was in truth a philosopher of pes-
simism; for in accepting this world as the best of all possible
worlds, he and his followers were thereby accepting the status
quo. Voltaire, a fighter, would not. And thanks to his battles
with injustice of many kinds, he became known as the conscience of Europe.

Second, Voltaire saw Leibniz's philosophy as illusionary, shuf-
fling, and pretentious—the absolute antithesis of Newton's and
a mockery of what philosophy should be. It was, in a sense, a
counterfeit science. Recall Leibniz's suggestion that his calculus
might be able to unlock the secrets of human behavior. Voltaire
had fun with this idea.

The Eel Man

Such was the man with whom John Turberville Needham, a
cocky cleric and scientist, was to cross pens. Needham was definitely not a pushover, however. He had some weapons of his
own, including a powerful righteousness, a deep-seated belief
that his science was providing strong support for his own
Roman Catholic religion, training in polemics, a thick skin, and

a knowledge that some of his own experimental work had turned European science on its ear.

Needham had learned his polemics while training for the priesthood, and he was a fighter, like Voltaire. Perhaps his readiness for battle had something to do with his family background: He came from a family of Roman Catholics in England, who had refused to attend the services of the established Anglican Church. He was sent to school in France and was ordained a secular priest in 1737, at age 24. It was while directing a Catholic school from 1740 to 1743 that he got a taste of natural science, and by 1743, he had published his first scientific paper.

While the paper was mainly on a geological subject, he appended a section on "some microscopical discoveries I lately made."[6] The first discovery, on the mechanism of pollen, was to win him recognition in the world of botany. The second was to lead to his becoming known forever as *L'Anguillard,* "The Eel Man." He published a book on *New Microscopical Discoveries* in 1745, which sold well, and he went on to become the first Roman Catholic priest elected as a Fellow of the Royal Society (FRS). In 1761, he was elected a Fellow of the Society of Antiquaries of London, and in 1773, he was appointed the first director of the Royal Society of Belgium, where he helped introduce advanced laboratory techniques into biological science.

Voltaire's *Akakia* was actually an early sally in the war that developed between Voltaire and Needham: Though the "dissertation" was aimed at Maupertuis, Needham was targeted as well. In the satire, Voltaire had accused Maupertuis of creating "eels" from fine flour (in an experiment that is described later in this chapter). Voltaire also describes in the book a superb dish, pâté of eels. As usual, Voltaire was willing to bend the facts a bit, for it was Needham, not Maupertuis, who had produced "eels" from fine flour. Maupertuis had written in support of Needham's work, however, and that had been enough for Voltaire.

Generation, Spontaneous and Otherwise

Before visiting Voltaire and Needham on the battlefield, it would be wise to get some insight into what they were battling about. The mechanical approach to the study of the physical world—taken, for example, by Galileo and Newton—had successfully explained a wide variety of phenomena. Inanimate matter was thereby seen to be subject to natural law. Was it possible, scientists began to wonder, that living matter was subject to such laws as well? A particularly intriguing question was that of *generation,* the production of offspring from—what? Could the creation of an embryo be understood in scientific, rather than religious or metaphysical, terms?

Attempts to explain this most confusing of biological phenomena had resolved themselves into two major, opposing schools of thought. The *preformationist* idea, which dominated the scene through the first half of the century, suggested that all embryos existed, preformed though infinitesimally tiny, in either the egg or the sperm. Similarly, plants were thought to arise from preexisting miniature organisms hidden in the seed. All that happens in generation, the preformationists believed, is that previously invisible parts become large enough to be seen. The Swiss naturalist Charles Bonnet declared that preformation was "one of the greatest triumphs of the human mind over the senses,"[7] which shows clearly that preformation was at least as much a philosophical as a scientific theory. Therefore, it is no surprise to learn that it had roots dating back to Aristotle.

In one version of the theory, called "emboîtement" or "encasement," every embryo contained within it a huge number of other embryos, all waiting for their appropriate time to emerge. All embryos, in other words, had been created by God at the Creation. While it seems an incredible idea now, it was perfectly respectable then; Leibniz had been a firm supporter of *emboîtement.*

On the other side were the *epigenesists.* Maupertuis, for example, argued that the preformationists were not giving an answer but were simply setting the problem back to an earlier time. For God, he wondered, is there any real difference between one moment in time and another? He pointed out too that the offspring of a couple tend to have resemblances to both parents, and that the existence of hybrids also argued against preformationist ideas. He and other epigenesists argued that each embryo must be formed anew from other, disorganized matter.

For specifics, epigenesists reached for some of the ideas that had proved so successful in the fields of physics and astronomy. Maupertuis suggested that some form of attraction was involved. The idea was quickly challenged by the strongly religious zoologist Réaumur, who objected to its occult nature (sound familiar?) and to the fact that simple attraction could not provide the guidance the particles needed to come together satisfactorily. In answer, Maupertuis suggested that the particles themselves have some sort of inner intelligence—à la Leibniz's monads.

The famous eighteenth-century naturalist Buffon, whose multivolume *Histoire naturelle* was to be a standard text for years, came up with several suggestions for epigenesis, including a *moule interieur* (internal mold), special "penetrating forces," and a division of matter into organic and brute forms. Excess organic matter, beyond what was needed by the body for its own purposes, became the seminal materials of both parents. Buffon had described Needham's observations in his *Histoire* and had thereby brought them into prominence.

Finally, Needham himself offered an epigenetic theory involving some sort of vegetative force (e.g., conversion of plant into animal matter) as the source of all life's activities. This force came in two forms—one involving expansion, the other resistance—and the balance of the two was what produced the phenomena of life. Needham was a Leibnizian, and his two kinds of forces are reminiscent of Leibniz's motor force and force of inertia, which make up what Leibniz called the "vis viva."

All three of these epigenetic theories (Maupertuis's, Buffon's, and Needham's) were similar, in that they involved the idea of a force as a cause of development. This convergence was no coincidence. First, recall Newton's success with a single physical phenomenon, gravitation, in the world of inanimate matter. Second, Buffon had worked with both of the other men.

Of the three, only Needham had done significant laboratory observations in the area of embryology. Undertaken around 1747, these observations were a reasonable attempt to check out some of the claims of the preformationists. He enclosed boiled mutton gravy in a glass vessel, then sealed the opening with cork and mastic, a resinous cement. As a further precaution, he heated the vessel in hot ashes; the intent of this action was to kill any living things that had remained in the flask after the boiling and bottling.

On opening the flask after a few days, he examined the gravy and saw that "it swarmed with Life, and microscopical Animals of most Dimensions."[8] In experiments with moistened, tainted wheat, he found a similar result. Among the "microscopical animals" were some that he described as looking rather like eels.

In other words, Needham claimed to have seen *spontaneous generation,* the creation of life from nonliving matter. The possibility of spontaneous generation had long been held; in 1667, a well-known Flemish physician and scientist, Jan Baptista van Helmont, believed that anyone could make mice by mixing dirty rags with wheat. How could anyone doubt this? All that was needed was to put the two items together in an open container, wait an appropriate time, and sure enough, the mice would appear.

By the middle of the eighteenth century, however, the idea of spontaneous generation had become somewhat disreputable, and it was fairly well accepted that life came only from life, and from the same kinds of life. Mice did not come from wheat and sweaty underwear. Now Needham, with his mutton gravy and wheat, using what seemed to be solid experimental technique, had turned the whole thing around again. Perhaps at this lower level of life, inorganic matter was indeed changed into organized, living creatures.

In any case, Needham's work did seem to spell death for preformation theory; certainly, it showed that there was no need for that concept. For even though Needham's experiment dealt with lower forms of life, the implication was that it could also be true of higher forms; so there was no need to have the embryos in existence from the original moment of God's creation.

To his dismay, however, his observations were seized by materialist and atheist philosophers, who found in them support for their own ideas. The materialists, for example, believed that everything could be explained in terms of matter in motion, or of matter and energy, and Needham's finding fitted beautifully with this belief. Also, if unorganized matter could form into living things, what need was there for a divine creator?

Voltaire was as upset by this outcome as Needham but, of course, blamed Needham for it. Though Voltaire fought against many of the excesses of the church, he was a firm believer in God—and he was a preformationist (as was Leibniz). Voltaire was also, as we know, a strong supporter of Newton, while Needham, the professed Leibnizian, believed in a vegetative force within every monad.

Ironically, just as Voltaire had learned of Newton's work through Maupertuis, so too did he first learn of Needham's work through that unfortunate man, this time from Maupertuis's letters, in 1752. It was not until the following decade, however, that Voltaire really began to go after Needham. A number of events conspired to make him catch fire. One was undoubtedly a long, impressive-looking book produced by Needham in 1750. While it included information on his observations, it was basically a rambling mixture of science, philosophy, and religious polemic, just the sort of thing that would drive Voltaire to distraction.

Goliath versus Goliath

Voltaire had moved to Geneva in 1755, where he would be somewhat less subject to censorship than he had been in France. But the Protestant Church was still a powerful force in Switzerland, and the subject of religious miracles began to get mixed up in politics. Jean Jacques Rousseau, another major French writer and polemicist, had written against miracles, in hopes of weakening the Roman Catholic Church's power. Needham defended not only miracles, but also both the Calvinist Church and the Roman Catholic Church, as well as the politics of the leading classes of the area.

In 1765, a series of pamphlets began to appear, anonymously but only barely so, titled *Lettres sur les miracles*. They were Voltaire's of course; he was taking aim not only at revelation miracles but also, as Hobbes had done a century earlier, at the divine right of kings. Because Needham's work appeared to provide clear evidence that miracles took place every day, Voltaire felt he needed to shoot Needham down. Voltaire suggested that Needham was a homosexual: "What! I shout, a Jesuit transfigured among us, a teacher of young men! This is dangerous in every way."[9]

Needham decided to do battle. In answer, he wrote several public letters to Voltaire. In one, he denounced Voltaire's libertinism, referring scornfully to "so-called sages" who rigorously profess, but do not practice, celibacy. He was referring here to Voltaire's several love affairs, the latest with Voltaire's own niece. Then he added that Voltaire's "writings are poison" and are "a public invitation to libertinage, which is the greatest threat to the population."[10]

According to Needham's letters, Voltaire, who pretended to be a great benefactor, was really the scourge of humanity and should be declared an enemy of the country. Needham wrote, "According to you, morality is a very slight thing and ought to be subjected to physics. I say that physics ought to be subjected to morality."[11]

The first two of Needham's letters were relatively straightforward; the third was a parody of Voltaire's third letter, and, he was sure, he had exposed Voltaire's "false reasoning." Convinced of his triumph, he wrote happily to a colleague, Charles Bonnet (ironically, a preformationist like Voltaire), that he had not written earlier because he was "finishing a small war against Mr. Voltaire." Referring to his letters as his "trophies," he added modestly that he had worked not for glory, but for the good of the society.[12]

Needham's triumph was short-lived. He had forgotten with whom he was dealing. Although Voltaire could laugh at himself (as he aged, he made fun of his own physical appearance—his toothlessness, his pipe-stem legs and skeleton-like figure), he was not about to have anyone else, especially Needham, laughing at him. Remember that while Voltaire was a mere "popularizer," Needham was by this time an important figure in science. A contemporary study of the articles in the *Journal des Savants* for that period finds Needham the most cited author.[13] How that must have aggravated Voltaire.

In a manner reminiscent of his battle with Maupertuis, Voltaire proceeded to change Needham's identity. He came up with the idea of a fanatic Irish Jesuit who wanted nothing more than to convert the entire Protestant world to Catholicism. Voltaire again used the idea of disguise: Needham, who in actuality did not wear priestly garb, now became a priest disguised as a human, one who was able to miraculously create "eels" out of mutton gravy and blighted wheat.

Needham, of course, was neither Irish nor a Jesuit, but by the time Voltaire got finished with him, he was both—at least in the eyes of Voltaire's readers. In that day, it was no compliment to be called either. In fact, the Jesuits had been expelled from France in 1764. Why Irish? Perhaps because as an Irish Catholic, Needham would no longer be an underdog in his own country and would thereby be more likely to appear as dangerous to Protestants.[14]

In his twentieth and final letter, Voltaire created a farcical scene in which Needham is jailed both for being a Jesuit and for hiding the fact. That, suggested the letter, was the end of Needham. The great success of the letters was proclaimed by the fact that in 1771, they were placed on the Catholic Church's *Index librorum prohibitorum.*

Needham refused to admit defeat, however. He issued an anonymous pamphlet defending the divine right of kings and arguing that Voltaire, a native of France, should not meddle in Genevan affairs. Arrows were being directed at Voltaire from other quarters as well. Buffon, never directly attacked by Voltaire, was nevertheless indirectly a target of his ridicule for having called Needham a good observer and for having collaborated with him. Buffon gave some of the fire back, however, suggesting that Voltaire's jealousy of every celebrity "has intensified his bile, which has been cooked with age. So that he seems to have formed a project intending to bury all of his contemporaries while they are living."[15]

Voltaire was particularly unhappy that Needham's work was widely cited and followed. He felt that a system proposed by the scientist d'Holbach was founded on Needham's observations, and in a letter to Suzanne Necker, dated 1770, he wrote, "It is shameful to our nation that so many people quickly embrace so ridiculous an opinion. One must be foolish not to admit a grand intelligence when one has one that is so small." Now note how he strengthens his invention when he refers to "a system based entirely on the false experiments made by an Irish Jesuit who has been taken for a philosopher."[16]

Relentless Pursuit

Note, too, the phrase "false experiments." Could this be a clue to his dogged pursuit of Needham? One scholar, Jean A. Perkins, argues that Voltaire had a deep fear of charlatanism and "was convinced that Needham had set up a fraudulent experiment in order to prove matter capable of organizing itself." Considering Voltaire's passionate pursuit of justice and his view of himself as a knight on a white charger, it is likely that this was an important reason for his vitriolic attacks.

One of the problems, as in the circle-squaring fracas a century earlier, was that biological concepts were so confused at the time that Needham could honestly feel that his experiments did show the reality of spontaneous generation and epigenesis, and there would be no way for anyone to prove otherwise. Needham was probably guilty of what might be called "selective perception," a common enough fault of researchers even today: One sees what one wants to see when there is in mind a preconceived notion.

We have already seen how Needham's ideas were taken up by atheists and materialists, however, and that Voltaire was deeply disturbed by this. Voltaire may have simply figured that the only way to deal with the situation was to destroy Needham as he had Maupertuis. In other words, while he excoriated others for their fanaticism, it is conceivable that he fell prey to a similar malady when it came to religious ideas. In fact, even their basic disagreement had religious overtones. Voltaire favored the concept of continuous creation mainly because he, like his idol Newton before him, could not accept the idea of a God who has nothing to do through the ages.

Though he had done some simple biological experiments with snails himself (and had reached an incorrect conclusion), he probably never really understood Needham's work. Still, he was right on the mark when he wrote, "One is free not to believe that the right toe attracts the left toe, nor that the hand places itself at the bottom of the arm by attraction"[17]—obviously a slap at Maupertuis's contribution to Needham's work.

Regarding Voltaire's feelings about Needham, his instincts were good here, too, for Needham had made a technical error in his experiments, which resulted in inaccurate observations. Hence, any systems based on them were obviously also incorrect. In fact, by 1765, Voltaire could call on some new work by Lazzaro Spallanzani, who was, like Needham, a cleric.

Whereas Voltaire could only attack Needham personally, Spallanzani, one of the greatest of all experimentalists, could go after Needham's science. To Spallanzani, this seemed necessary if the atheistic materialists' use of Needham's work was to be undercut. Spallanzani showed, finally, that where Needham thought he had closed his flasks so well that microscopic life could not enter, his cork stoppers were not up to the task. Spallanzani sealed his flasks by melting the glass itself. He also showed that the use of hot ashes to heat the flasks was not sufficient to kill the microscopic life in them, and that boiling for at least three quarters of an hour was necessary. Needham, in other words, had neither completely killed off the organisms that were in the flasks nor prevented new ones from entering after the flasks had cooled.

Needham, not satisfied that he had truly been refuted, argued that Spallanzani's intense boiling had destroyed not only the germs but also the germinative power, or vegetative force, of the mixture. Spallanzani easily dealt with this objection by showing that when the infusions were again exposed to the air, no matter how intense or prolonged the boiling to which they might have been subjected, the microscopic life again appeared.

Voltaire was, of course, delighted with this new development. In a letter to Spallanzani, he wrote, "You have given the final blow, Monsieur, to the eels of the Jesuit Needham. They have wriggled very well but they are now dead. . . . Animals born without a seed cannot live a long time. It is your book that will live, because it is founded on experiment and on reason."[18]

Note the peculiar irony here. Needham's observations were indeed wrong, yet he was supporting what turned out eventually to be the winning side. Spallanzani, on the other hand, was in truth a careful and accomplished experimenter, but he incorrectly concluded that he had found proof for preformation, which turned out to be a wholly wrong theory.[19] Worse, over the years, belief in this theory seriously delayed the development of embryology.

In 1759 and, more finally, in 1768, Kaspar Friedrich Wolff advanced the cause of epigenesis by concentrating on certain parts of an organism, such as blood vessels in developing chick embryos, and showing that they do indeed develop from a different kind of tissue. His work, oddly enough, was inspired in a double sense by that of Leibniz—first, by Leibniz's calculus, which is, after all, a mathematics of change; and second, by Leibniz's notion of monads.

Nevertheless, a fervently believed idea, even if wrong, dies hard. As long as some proponents are alive and unconvinced, the idea has life as well. The longevity of wrong-headed ideas is repeated several times in later chapters of this book. Thus, even in the 1770s and 1780s, preformation was still going strong. In 1776, Needham issued his *Idée sommaire* against Voltaire and was still railing against the "numerous absurdities" of the preformationists.[20]

It took later research by many workers—work involving the cell theory of life and, later, chromosomes—before the final death knell could be tolled for the tenacious doctrine. Thanks to this work, we now have excellent descriptions of an amazing variety of developmental histories, all showing a common pattern of development.

Notwithstanding the somewhat foolish aspects of both Voltaire's and Needham's stands, both men were highly honored in their time. Voltaire was able to pull Needham down several pegs, but Needham proved far more resilient than Maupertuis. In fact, by 1781, when Needham died in Brussels at age 68, he had been honored with English and Belgian titles of nobility and a number of ecclesiastical titles as well.

Voltaire, who had seen Newton's funeral when a resident in England, decided then that he would like to have one as impressive. He got his wish, though it took 13 years after his death for it to happen. When he died of a fever in 1778, he was buried in haste outside Paris, to assure him a Christian burial. Then, by popular demand, his remains were transferred in 1791, with great pomp and ceremony, to the newly completed Pantheon in Paris.

The Final Solution

How are we to look upon Voltaire and Needham today? Voltaire, like so many others in his day, got his religion mixed up with his science. It was, as we have seen, at least partly because of the atheists' use of Needham's work that Voltaire was so worked up. Yet what were Voltaire's religious beliefs? Basically, although he rejected organized religion, he felt that there was order and harmony in the universe, which testified to the existence of an intelligence as a basic mover. As he put it, a watch demands a watchmaker. This idea is still widely held today.

The Needham–Voltaire feud also involved the differences between a static and a changing, or evolutionary, universe. Again, the discussion helped establish a foundation on which later workers could build. Suffice it to say here that Voltaire opted for a static universe; he argued, as any preformationist had to, that the world was now as it was created at the beginning. In answer to questions about marine fossils found in the Alps, he suggested that they were the remains of meals taken by travelers passing by.[21]

Regarding Needham: Was he simply blinded by his religious fanaticism? Rachel Westbrook, a historian of science who did one of the few full-length studies of Needham, suggests that he "stands out as the last of a vanishing breed"—that is, one who used science to defend his religion. Yet, she adds, it is ironic that "much of his thought contributes to the new, secular view." There is, for example, a "dynamic" in his system and an emphasis on nature in flux. "God or spirit as an explanatory necessity," she adds, "is almost superfluous in Needham's system of nature."[22]

Today, it is obvious that development is epigenetic. There clearly is no preformed organism in either egg or sperm. Nonetheless, still today, one can bring forth Needham's argument that God played just as important a role in Needham's theory as in that of the preformationists. What difference did it make, he wondered, if God made all organisms at the beginning of the world, or if he simply established the laws through which all future life would arise?

In other words, unless the creation of living things is assigned to a divine creator, then somehow life did arise from nonlife, and the concept of spontaneous generation has not actually been buried, but rather moved back to an earlier time. Some current work suggests that if this did take place in a remote era, it was at the level of the virus or below.

Hence, although the preformation/epigenesis argument has been won by the epigenesists, the vitalist/mechanist debate, of which it was a part, continues. The *vitalist* position—as expounded by Leibniz, for example—is that the particles of living matter are somehow different, intrinsically, from those of nonliving matter. The *mechanists* argue that matter is matter, and that the phenomena of life can be explained in terms of how the particles are put together.

Which side is right? We are not a great deal further ahead in making that decision than were the scientists and philosophers of Leibniz's day. While treating living things as if they are machines has been extremely useful in research, it may be that the more metaphysical vitalistic approach will be needed to carry us to a further understanding of life. At the very least, Needham's work suggested that somehow life's activities came from within, rather than from without, which was a pretty good way to start. In addition, the idea of an internal mold (the *moule interieur*) is not a bad first description of the work performed by DNA.

For Voltaire, the dispute with Needham was just one of many. His general feeling about disputes can be summed up with his comment, "Disputes among authors are of use to literature; as the quarrels of the great, and the clamours of the little, in a free government, are necessary to liberty."[23] While Voltaire did not mention science in this comment, the argument holds there, too. Rephrased, it might read, "Disputes among natural philoso-phers are of use to science, as the quarrels of the great, and the clamors of the little, are necessary to freedom of thought and the advancement of learning."

Preformation theory has been taken to task for holding back the progress of developmental biology. In 1931, the great histo-rian of science George Sarton wrote, "Thus was the fine obser-vational tradition of the seventeenth century interrupted, or at any rate considerably slowed down for more than a century, by discussions which were irrelevant, because they were too far ahead of the experimental data."[24] It can be argued that the dispute between Needham and Voltaire was an important fac-tor in bringing the spontaneous generation question to a head, and that it thereby caused the experimental evidence to appear that finally put the work back in gear.

Today, the study of development involves much more than re-production of offspring, of course, and includes aging, the pos-sibility of rejuvenescence, and even cancer. Suppose, for exam-ple, that cancer is found to be caused by viruses or some even lower life form, created de novo. It's not likely, but it could hap-pen. Needham would laugh. Voltaire would smile and yell to his secretary, "Quickly, Wagnière, a pamphlet."

CHAPTER 5

Darwin's Bulldog versus Soapy Sam

Evolution Wars

Part 1: The Nineteenth Century

The hall at Oxford University was packed on that Saturday afternoon in the summer of 1860. Well over 700 bodies were stuffed in, with the center section a solid mass of clerical black. Scattered around the rest of the hall were a few defenders of Charles Darwin's newfangled theories. The occasion was the annual meeting of the British Association for the Advancement of Science; the date was June 30, roughly seven months after the publication of Darwin's provocative new book, *The Origin of Species by Means of Natural Selection.*

At the moment, John William Draper, temporarily imported from New York University, was droning on about "The Intellectual Development of Europe Considered with Respect to the Views of Mr. Darwin"—actually a talk on Darwinism and social progress. Already the broad implications of the theory were being recognized and discussed.

But everyone knew that the illustrious bishop of Oxford, Samuel Wilberforce, was planning a massive ecclesiastical attack on Darwin's dangerous new ideas.

Darwin himself was home sick, but he would have been of little use in any case. Born in 1809, he was just 50 years old when his *Origin* was published. By this time he had changed from a vigorous young explorer to a reclusive, sickly stay-at-home. Though his ideas on evolution were considered outrageous, he himself was inordinately shy, and would never have been able to stand up to such a rhetorician as Bishop Wilberforce.

The nickname "Soapy Sam," given to Wilberforce by the students at Oxford, is often taken today as strictly pejorative. At the time, however, there was mixed in a real respect for his oratorical prowess, especially his ability to inject charm and wit, along with venom, when necessary. He also had some standing as a mathematician. Further, although Wilberforce was not a scientist, he had been carefully primed for the attack on Darwin by Sir Richard Owen, recognized as the leading comparative anatomist of his time.

Although Darwin's work had been published only months earlier, it had already created a considerable stir. One reason is that he was already a recognized and respected naturalist; in fact, he had been proposed for knighthood shortly before his book's appearance. Prince Albert had already agreed to the proposal. With the appearance of the *Origin,* however, Queen Victoria's ecclesiastical advisers—including Bishop Wilberforce—argued against the proposal, and it was defeated.

Darwin, it should be noted, was not the first to come up with a theory of evolution. The idea that species are not immutable but can change and adapt over the course of time had been put forth innumerable times. Darwin's own grandfather, Erasmus Darwin, had also propounded the idea, as had Lamarck (who believed that changes caused by exposure to environmental influences could be passed on to descendants).

Nonetheless, in Darwin's day, far more people were violently disposed against any form of evolution than were for it. An earlier proevolution tract, *Vestiges of the Natural History of Creation,* had been published in 1844 by Robert Chambers, a successful science popularizer. Expecting a riotous reception, he had not put his name to the work.

Riotous indeed: Adam Sedgwick, Darwin's geology teacher at Cambridge, growled that evolution and spontaneous generation coupled in an "unlawful marriage" and spawned a hideous monster; it would be merciful to crush "the head of the filthy abortion, and put an end to its crawlings."[1] Sedgwick, after all, believed that a proper balance between living things and the world in which they live was maintained by divine intervention whenever needed. Darwin was destroying this link, not to mention the link between the moral and the material worlds.

Chambers's book was also badly mauled by Wilberforce himself in the influential *Quarterly Review;* for, unfortunately, *Vestiges* was, in truth, a pastiche of sense and nonsense. Darwin himself had viewed the book with mixed feelings. While he had some problems with the basic science behind it, especially in earlier editions, he also felt that it was perhaps drawing some fire away from the expected reaction to his own long-gestating treatise and might even be preparing some ground for it.

The attacks on evolutionary ideas, while excessive, were not entirely unreasonable. Essentially, evolutionists argued against the need for innumerable special acts of creation, meaning a separate one for every species of life on Earth. In *Vestiges,* however, as well as in all the other earlier attempts, these ideas were pure speculation, with little to back them up, and utterly lacking any mechanism that might explain how evolution happens. The creationist idea was as good as any.

This missing mechanism—natural selection—is what Darwin added, along with a huge mass of supporting data and a well-reasoned argument. Here, briefly, is Darwin's own description of natural selection:

> Let it also be borne in mind how infinitely complex and close-fitting are the mutual relations of all organic beings to each other and to their physical conditions of life; and consequently what infinitely varied diversities of structure might be of use to each being under changing conditions of life. . . . If such [variations] do occur, can we doubt (remembering that many more individuals are born than can possibly survive) that individuals having any advantage, however slight, over others, would have the best chance of surviving and of procreating their kind? . . . This preservation of favourable individual differences and variations, and the destruction of those which are injurious, I have called Natural Selection, or the Survival of the Fittest.[2]

This hardly sounds threatening to (most of) our ears, but it was this aspect that was quickly seen as a real danger by the clergy. They therefore felt it was important to demolish the whole Darwinian structure quickly. For them, the 1860 meeting of the British Association for the Advancement of Science (BAAS) seemed to be an ideal venue.

Darwin's defenders, small in number, were hoping that someone would stand up against Wilberforce. Among these defenders was Thomas Henry Huxley, a highly respected scientist who had made important contributions in zoology, geology, and even anthropology. He also wrote on education and religion in a clear and beautiful style and was an accomplished speaker.

Huxley had an additional reason for supporting Darwin: As he put it to a colleague, cleric Charles Kingsley, "It is clear to me that if that great and powerful instrument for good or evil, the Church of England, is to be saved from being shivered into fragments by the advancing tide of science—an event I should be very sorry to witness, but which will infallibly occur if men like Samuel of Oxford are to have the guidance of her destinies; it must be by the efforts of men who, like yourself, see your way to the combination of the practice of the Church with the spirit of science."[3] In other words, evolutionary science and religion can coexist, but not amid extremist bombardment.

One of the problems with Darwin's ideas surfaced early. The basic concept, modification of species by natural selection, was at the same time so simple that Huxley could say, "Why didn't I think of that?" and yet so profound that it could mean many different things to different people, including his supporters as well as his opponents—so profound, in fact, that generations since have still not been able to comprehend its grand plan.

Another of the difficulties is its two-part nature: first, evolution itself, and second, natural selection, which seems to be the major stumbling block. The main problem with natural selection was, and still is, that there is no active selector in the "selection" process. Rather it is an a posteriori process: Nature does the selecting. Perhaps "natural preservation" might have been an easier term to accept.

According to Ernst Mayr, a present-day bulldog of high rank, Huxley himself never believed in Darwin's process of natural selection. Others also argued against natural selection and offered a variety of their own alternatives to explain the mechanism of variation. With one exception, none has stuck. The exception is *saltationism,* which involves jumps in evolution. Huxley himself questioned Darwin's insistence on *gradualism,* small changes that eventually add up to the great differences among species. In this, Huxley may have been right. In our own time, the much respected evolutionary biologist Stephen Jay Gould and his colleague Niles Eldridge have offered up their own version of the saltation idea, which they call "punctuated equilibrium." Gould is careful to point out, however, that this saltationism in no way negates the basic soundness of natural selection.[4] In any case, all the evolutionists, whatever their stripe, were firm in arguing against special creation.

In the Arena

Huxley, knowing of Darwin's understandable nervousness about his work, had written to him the day before *Origin* was published to offer support and encouragement: "And as to the curs which will bark and yelp, you must recollect that some of your friends, at any rate, are endowed with an amount of combativeness which (though you have often and justly rebuked it) may stand you in good stead. I am sharpening up my claws and beak in readiness."[5]

Although Huxley knew that Wilberforce was "a first-class controversialist," Huxley had attained a reputation as a solid debater and lecturer himself. In February of 1860, however, Huxley had lectured on Darwin's ideas at the prestigious Royal Institution[6]—with two surprising results. First, he "disappointed & displeased" everybody by trying to present all sides of the picture. Second, while looking to pry science from ecclesiastical control, he found himself taking a distinctly confrontational attitude toward the church-based side—much more so than Darwin had taken.[7]

Knowing that the Oxford session would be packed with clergy, Huxley decided early on not to attend. As he wrote later, "I was quite aware that if he [Wilberforce] played his cards properly, we should have little chance, with such an audience, of making efficient defence."[8] On the day previous to the session, however, Huxley had, by chance, met Robert Chambers, author of *Vestiges,* the earlier tract on evolution. He told Chambers of his decision not to attend, that he did not "see the good of giving up peace and quietness to be episcopally pounded."[9] Chambers, perhaps hoping for some long-delayed revenge against the man who had savaged his book, somehow managed to persuade Huxley to make an appearance and present an answer to Soapy Sam.

Thus, one of the great epics in the annals of scientific debate was set up. Unfortunately, the details are mired in mystery and confusion. The result, in fact, is a first-class *Rashomon,* with various reporters giving their own versions.

Despite the many variations, all who wrote about it later agreed on the general tone. Draper droned on for an hour until, finally, the floor was thrown open. Expectation ran high and, sure enough, Wilberforce rose to make a "few comments."

Fluent in the ways of high-class debate, Wilberforce began by briefly summarizing the common ground between science and the church—this was, after all, a scientific meeting. He even complimented Huxley, who, he was sure, was about to demolish him. Then he got down to business. Again, the exact words are in doubt, but in part, he accused the theory of being merely "an hypothesis, raised most unphilosophically to the dignity of a causal theory."[10] Similar objections are still advanced today.

After fully half an hour of wandering rhetoric, he pointed out how extremely uncomfortable he would feel if someone could show that an ape in the zoo was his ancestor (something that Darwin had never said or thought). He then turned to Huxley and asked slyly whether it was through his grandfather or his grandmother that he claimed his descent from a monkey. The audience exploded with laughter and cheers. Huxley merely murmered: "The Lord has delivered him to my hands."

This was a pre-electronics era, however, and Huxley knew he had not the blaring voice of a Wilberforce, so he held off on replying until the audience began to call, "Huxley, Huxley." Only then did he stand and offer his brief reply. "I am here only in the interest of science, and I have not heard anything which can prejudice the case of my august client." After a few more remarks in defense of the Darwinian point of view, he concluded, "Lastly, as to the descent from a monkey, I should feel it no shame to have risen from such an origin. But I *should* feel it a shame to have sprung from one who prostituted the gifts of culture and of eloquence to the service of prejudice and of falsehood."[11]

Keep in mind that insulting a bishop in those days was no light matter. The response was predictable: The clergy roared their outrage; Darwin's supporters cheered; and the students undoubtedly cheered both sides. One woman, Lady Brewster, fainted in shock.

There was more. Sir John Lubbock, a well-known astronomer and natural scientist, rose and made some remarks in support of Darwin's ideas. In contrast, Robert Fitzroy, with whom Darwin had sailed during the epic five-year voyage that had probably started the whole thing, and who was now an admiral and the ex-governor of New Zealand, stood up and, flourishing his Bible, claimed it to be the source of all truth.

Then the well-known British taxonomist and botanist Joseph Dalton Hooker added a powerful closing. After stating that Wilberforce obviously had not read Darwin's book and was equally obviously ignorant of the rudiments of botanical science, he added the clincher: "I knew of this theory fifteen years ago. I was then entirely opposed to it; . . . but since then I have devoted myself unremittingly to natural history; in its pursuit I have traveled round the world. Facts in this science which before were inexplicable to me became, one by one, explained by this theory, and conviction has been thus gradually forced upon an unwilling convert."[12]

These may have been the last words at the battle, but they were hardly the last words in the war. A few days after the debate, Wilberforce's review of the *Origin* appeared in the prestigious *Quarterly Review*. In this massive 17,000-word critique, he stood on firmer ground. He also attacked most strongly in an area that Darwin had hoped to sidestep: In the entire volume, Darwin had assiduously stayed away from that hot potato called "man." The closest he came was to predict, in his closing remarks, that with the help of additional, "far more important researches. . . . Much light will be thrown on the origin of man and his history."[13]

Wilberforce, however, knew this was a sore spot and went for it. He claimed that Darwin had applied his scheme of natural selection not only to animals but also to human beings. This was too much. "Man's derived supremacy over the earth; man's power of articulate speech; man's gift of reason; man's freewill and responsibility; man's fall and man's redemption; the incarnation of the Eternal Son; the indwelling of the Eternal Spirit,— all are equally and utterly irreconcilable with the degrading notion of the brute origin of him who was created in the image of God."[14] In other words, say what you like about the rest of the animal kingdom, humans have been specially created—and within the past few thousand years—period.

In the following years, many of Darwin's more progressive contemporaries followed Hooker's lead and moved into the evolutionary camp, though once again, their moves had many meanings. Then, 12 years after *Origin*, Darwin followed up with his next entry, *The Descent of Man and Selection in Relation to Sex*, in which he did indeed indicate that natural selection applied to humans as well. Again, the feathers flew. According to Mayr, "no Darwinian idea was less acceptable to the Victorians than the derivation of man from a primitive ancestor. . . . The primate origin of man . . . immediately raised questions about the origin of mind and consciousness that are controversial to this day."[15] As I show below, this idea took the brunt of the attack; natural selection was rarely mentioned thereafter.[16]

Among the most boisterous voices, on both sides of the evolution argument, were the satirists and cartoonists who, in Victorian England, made up a journalistic class all their own. Referring to the ongoing battle between Huxley and Richard Owen (the comparative anatomist and paleontologist who had prepared Wilberforce for his 1860 speech), *Punch* predicted in May 1861:

> Then Huxley and Owen,
> With rivalry glowing,
> With pen and ink rush to the scratch;
> 'Tis Brain versus Brain,
> Till one of them's slain;
> By Jove! it will be a good match![17]

Pamphleteers, too, had a high time. In one pamphlet, Huxley and Owen were charged with calling each other awful names;

Huxley called Owen a "lying Orthognathus Brachycephalic Bimanous Pithecus." Owen charged that Huxley was "nothing but a thorough Archencephalic Primate."[18] Owen also objected that Huxley couldn't be sworn in as a witness because he did not believe in anything. (Huxley later introduced the term *agnostic* in describing his own philosophy.)

Science in general was lampooned as well. Another pamphlet referred to a paper called "On the Passage of Palarized [sic] Light through smoked glass, a brick wall, a sheet of four-inch armour plate, and a dark room." The signers included "A. B. Surd, A. L. Chemy, A. Vision Ary, and A. Muddle."

With photography in its very early stages, cartooning was a powerful and widespread method of illustration. A common theme was Darwin as monkey or ape. One cartoon showed an approving monkey reading a Darwin book. In another, an ape-like Darwin is seen taking the pulse of a woman.

Literary people got into the act too. Novelist and polemicist Samuel Butler, for example, wrote several works satirizing and attacking Darwin's ideas. Examples include *Life and Habit* (1877) and *Evolution Old and New* (1879).

Religion

Darwin's reaction to the many attacks was mainly anguish. His biographers Adrian Desmond and James Moore subtitled their 1991 biography, "The Life of a Tormented Evolutionist." Darwin's main concern was the distress of his beloved wife, who had great difficulty in trying to reconcile her deep religious faith with her love and respect for her husband. The more the religious establishment attacked him, the greater was her distress.

Darwin's true feelings about the origin of species are not clear. In the first edition of the *Origin*, for example, there is simply no mention of a "Creator." In the closing lines, he referred to "this view of life, with its several powers, having been initially breathed into a few forms or into one." By the second edition, which appeared shortly after the first, he had changed the line to read, "having been initially breathed *by the Creator* into a few forms or into one" (emphasis added).

No one can say for sure whether the omission in the first edition was an oversight—not very likely, considering his careful nature—or whether the addition to the second edition was his attempt to assuage the distress he was causing his many colleagues and friends, as well as his wife. It is on this point that a big—and important—split occurs. Many religiously oriented people, lay and cleric alike, have no problem with accepting the basic ideas of evolution and even of natural selection—as long as they can continue to believe that God is in there somewhere. The most logical point for God is right at the beginning. Recalling the watchmaker question of Newton's day, the question then became, "Was it only necessary to set the species going, after which everything took care of itself, or was periodic intervention necessary in order for smooth operation to occur?"

The *Origin,* of course, had to stand or fall on its own merits, not on the debating abilities of Huxley or of any other defenders of Darwin. Fortunately, the book is a true masterpiece and has held up well. In the following years, its solid set of observations, its admission of holes where these existed, and its readability had a powerful effect. Geologists, first, then biologists, paleontologists, and others—both in the world of science and outside it—began to swing over into the Darwinian court. The swing, however, was slow, uneven, and by no means complete.

Objections

Not all objections were religious ones. One of the major stumbling blocks came to the fore at the next annual meeting of the BAAS, when William Thomson, later to become Lord Kelvin, presented his calculations on the age of Earth. The limits placed by him—roughly 100 million years—were simply too tight to allow for the time needed by natural selection to do its work.

By the time Darwin published his fifth edition of *Origin,* he considered Kelvin's calculations a real problem and tried to deal with them. In the sixth and last edition (1872), he admitted that the objection was "probably one of the gravest as yet advanced." Yet, in his careful, perceptive way, he added, "I can only say, firstly, that we do not know at what rate species change as measured by years, and secondly, that many philosophers are not as yet willing to admit that we know enough of the constitution of the universe and of the interior of our globe to speculate with safety on its past duration."[19] As the next chapter shows, it took another 40 years before Kelvin's objection tumbled.

In the same paragraph in *Origin*, Darwin dealt with another admitted problem, "the absence of strata rich in fossils beneath the Cambrian formation." He pointed out with astonishing perception that "although our continents and oceans have endured for an enormous period of time in nearly their present relative positions, we have no reason to assume that this has always been the case; consequently formations much older than any now known may lie buried beneath the great oceans."

Summing up his treatment of the various objections, he added, "I have felt these difficulties far too heavily during many years to doubt their weight. But it deserves especial notice that the more important objections relate to questions on which we are confessedly ignorant; nor do we know how ignorant we are."

A major hole, again well recognized by Darwin, was the mechanism by which variation and modification took place. The beginnings of an answer were provided by Johann Gregor Mendel, an Austrian monk and experimental botanist whose famous experiments with pea plants, conducted in Brno, Czechoslovakia, laid the foundations for genetics. Daniel C. Dennett, another of today's Darwinian bulldogs, maintains that a copy of Mendel's famous paper lay, unread, in Darwin's study during the late 1860s.[20] Science historians generally agree, however, that Darwin was not aware of its implications and that it played no part in his work. Although Mendel's paper had been published in 1865, six years before Darwin's *Descent of Man*, it had appeared in an obscure Czech journal. One has to wonder whether and how things might have been different had Darwin read Mendel's paper.

In any case, Darwin had treated genetic variation as an unknown factor. Familiar with artificial selection in animal breeding, he knew that variation certainly existed, and he built on that. The problem that Mendel's work solved was this: Darwin assumed that the action of selection is slow. It was also commonly believed that under continued mating among the individuals in a species, the variants, whatever they may be, merely blended back into intermediacy. If variations disappeared, however, how could natural selection do its job? Mendel's laws of inheritance showed that the changing characteristics did *not* blend, but remained distinct, so natural selection could operate slowly and still do its work.

Part 2: The Twentieth Century

Mendel's obscure paper was rediscovered in 1900 by Hugo de Vries, a Dutch plant physiologist. De Vries then went on to add his own theory of mutation, which involved sudden changes in the germ plasm. He also had a brilliant intuition that had to do with X rays, which had just been discovered. He guessed that because these rays can penetrate living tissue, they might very well be able to alter the hereditary particles, whatever they may be, in germ cells.

This speculation, however, was not to be demonstrated for another two decades. Then, starting around 1919, the American geneticist Hermann Joseph Müller showed that the genetic material could indeed be affected by the environment, and he proved that the changes could then be passed on to the organism's descendants. With these discoveries, much of what was missing fell into place, and Darwin's theory became far better rounded. Now, finally, the evolutionists thought, who can doubt the truth of their stand?

Had evolution and Darwinism indeed triumphed? Hardly. In fact, writes Mayr, "From about 1890 to 1910, Darwin's theory was threatened to such an extent by various opposing theories that it was in danger of going under."[21] Even after Mendel's paper was (re)discovered, it took a long while for the implications of the genetics work to sink in and spread. In fact, as with all new and important developments, there was even some negative effect on the basic theory, with geneticists pitted against strictly Darwinian evolutionists.[22]

By the 1920s, a solid foundation had been laid for a new understanding and appreciation of evolution, and this new theoretical edifice took its place, at least in the scientific world, as a scientific fact. As was the case with Newton's laws, there might be some advances and emendations, but there was already widespread support for evolutionary theory.

The theory was also being taught in some of the country's schools, including many of the new high schools springing up across the country. Ronald L. Numbers, in his excellent history of creationism, has argued that some of this teaching might have been too strident, too willing to pit evolution against religion. This boldness, he feels, might have elicited a sort of counterrevolution among a group that just might have learned to accept evolution.

He writes, "Even in the theologically conservative South, a number of church-related colleges had been teaching the theory of evolution for decades." But in the years before World War I, for whatever reasons, angry antievolutionists increasingly "identified organic evolution as the cause of the social ills plaguing modern civilization."[23] The consequences of this identification have been profound.

The Monkey Trial

Darwin on the Ropes, one of the great (anti)morality plays of the Western world, has been seen just twice on major stages. The first production, described earlier in this chapter, took place at Oxford University and was seen by fewer than a thousand people, but it reverberated in print for decades. The next production came about in a different way.

While the evidence and support for evolution grew over the following years, the antievolutionary forces were digging in their heels, especially in the United States. With no established church to lead the crusade against evolution, independent religious sects had been springing up with electrifying speed and diversity, and with teachings that ranged from sensible to ridiculous. One common factor, however, was a belief in the revealed truth of the Bible's teachings. Strong believers in the specifics were, and still are, called "fundamentalists."****

The part of the Bible that seemed most directly contradicted by Darwin's evolution is the opening story of Genesis, particularly the description of creation. Defenders of this story, who were convinced that it described exactly the beginnings of all life on Earth (as well as all of our observed universe), took on or were given the name "creationists," and in some regions of the country, they wielded considerable power. By the early 1920s, they were able to get the teaching of evolution outlawed in three American states: Tennessee, Mississippi, and Arkansas.[24]

Evolutionists, aghast, were anxious to have the question debated in a court of law. So it was that the next major production of *Darwin on the Ropes* took place, 65 years after the first one, in a sleepy little town in Tennessee. In this 1925 production, John Thomas Scopes, a young high school science teacher and football coach, was tried for teaching the theory of evolution, in violation of state law. Staged by a small group of interested parties, including the American Civil Liberties Union, this production was far more elaborate than its predecessor.

Stretching over weeks and observed by dozens of journalists, the trial was followed intently via telegraph and newspaper all over the world. Among the journalists sat H. L. Mencken, possibly the most influential American essayist and social critic of the time. This trial gave him plenty of opportunity to show his biting, satirical style.

Although this production is sometimes subtitled "The Scopes Monkey Trial," Scopes himself had a very small part. Again, there were two leading roles: The first, Wilberforce's part, was played by congressional representative and three-time presidential nominee William Jennings Bryan, a fiery orator and political evangelist with roots deep in the soil. With a national reputation and a powerful animus against evolution (Darwin-

----

****Because fundamentalism as a movement is not a monolithic edifice, and has even had some liberal tendencies, the tendency being discussed in this section might better be called biblical literalism. "Fundamentalism" is more common, however, and will be used here.

ism, he was sure, lay at the heart of World War I), he was perfect for the part. In fact, even Mencken, who had no love for Bryan, said that as a rhetorician, Bryan was "the greatest of them all."[25]

Huxley's role was taken by the urbane, agnostic, highly successful criminal trial lawyer Clarence Seward Darrow. Darrow had so little love for Bryan and his ideas that he not only volunteered his services but also waived his usual high fees and even paid his own expenses.

Oh yes, the judge: He was John T. Raulston, who sat under a banner that stated, "Read Your Bible Daily." Of the 12 jurors, 11 were fundamentalists, 1 was illiterate, and not a single juror knew a thing about science or evolution.[26] Darrow's theme was, "After a while, your Honor, we will find ourselves marching backward to the glorious days of the 16th century when bigots lighted the fagots to burn men who dared to bring any intelligence and enlightenment to the human mind."[27] This went over big with the judge.

In fact, one of Raulston's first decisions was to prevent the defense from presenting expert witnesses. The prosecution's argument: We don't need them. Bryan, holding up the textbook Scopes had been using, showed a picture of a human along with other mammals and thundered, "How dare these scientists put man in a little ring with lions and tigers and everything that smells of the jungle. . . . One does not need to be an expert to know what the Bible says."

Dudley Field Malone, a member of Darrow's team, correctly answered, "I have never seen greater need for learning than is exhibited by the Prosecution."[28] To no avail. The judge did not permit the defense to call its expert witnesses. One commentator called this a wise decision, pointing out, "If a great state has decided by law that twice two are five, it would be foolish to allow mathematicians to testify."[29] Darrow had what the experts would have said in court typed up and distributed to the press. By the very fact that the testimony had been barred, it became more interesting to the outside world.

Darrow then invited Bryan to appear as an expert witness on the Bible *for the defense*. Bryan, in a fit of overconfidence, made a second tactical error: He accepted. The result—a Christian fed

to the lions—was so devastating that the judge had the evidence
struck from the record. Nonetheless, this change to the court
record did not prevent the verbal pummeling of Bryan from
streaking out to the far corners of the earth via the dozens of
journalists who ate up every word. Bryan's mistake proved a disaster for his side, at least in the court of world opinion.

But, although Bryan might have already lost in the world
court, Darrow was aware that Bryan was nevertheless a strong
performer in a familiar and sympathetic venue, and Darrow
had no wish to permit him to recoup his strength. Therefore,
Darrow forced the proceedings to a screeching halt by having
Scopes plead guilty. The judge fined Scopes $100, which turned
out to be another problem for the prosecution. Legally, it is the
jury that should decide the award, and the whole thing was
thrown out a few months later by the Tennessee Supreme Court.

As in the Oxford debate, the defenders of evolution felt the
fires of victory and looked forward to a future of unfettered
freedom for their cause. Alas, it was not to be. In fact, even
though the state court struck down the conviction, it upheld the
antievolution statute.

Constant Pressure

Although the evolutionist team in Dayton had accomplished exactly what it set out to do, and although Tennessee had come
out with egg on its face, the antievolutionary forces were completely undaunted. The truth is, it was not until 1967 that edu-
cators could legally teach evolution in Tennessee.[30] Further, the
continuing controversy had the effect of keeping evolution out
of many of the country's schools and school textbooks, particularly in the South. A dozen more antievolution bills were presented immediately following the Scopes trial, and two, in Mississippi and Arkansas, passed. Then, two major occurrences
added strength to the evolutionist cause.

During the early part of the twentieth century, most workers
in the broad field of evolution fell, roughly, into three separate
disciplines: genetics, systematics (taxonomic classification of
biological groups), and paleontology.[31] Each group had something of value for the others, but they were like different departments in a university, each having little use for the other.

Eventually the three disciplines began to mix and to strengthen each other, like sand, cement, and water. The foundations had been laid for a single structure, and it began, slowly, to rise. Julian Huxley, grandson of Thomas Henry Huxley and a prominent biologist himself, introduced the term "evolutionary synthesis" in 1942 to describe a unified theory in which *microevolution*—the genetic aspect—meshed nicely with *macroevolution*, which had to do with living, full-scale organisms. (Another term sometimes used is "new synthesis.")

Not long after, in 1957, the USSR caught Uncle Sam napping, with an initial success in the space race (*Sputnik*). This opening round of the contest galvanized the science establishment and led to a concerted effort to improve the teaching of science to American young people. Such teaching included recognition of evolutionary theory as an absolutely necessary foundation, a basic organizing principle, for modern biology.

The Arkansas statute banning the teaching of evolution had worked for a while but was finally carried all the way to the U.S. Supreme Court and was struck down in 1968; even though the Arkansas law did not specifically mention the biblical account of creation, the court recognized that the law had that intention.

At the same time, however, creationists had already come to the conclusion that their initial approach, religion-based creationism, was not going to work. It obviously flew in the face of the First and the Fourteenth Amendments to the U.S. Constitution.[32] The solution: Transform the doctrine into a "science." By using scientific phraseology, the creationists can thereby argue that their doctrine has as much right to be taught in schoolrooms as has evolution. Thus, instead of creationism, we now have "creation science." Later, when the word "creation" became suspect, the creationists changed the name of their belief system again, labeling it "Intelligent Design Theory."[33]

Using the new approach, they again were successful in getting laws passed here and there. What they hoped for, of course, was replacement of that "unproved hypothesis," evolution, with creation science. Mostly, the resulting laws called for the teaching of both approaches. In 1981, for example, Louisiana passed a law that required any public school teaching the theory of evolution to also teach creationism as a science. The

U.S. Supreme Court ruled 7 to 2 against it, again recognizing the intent of the law as a religious one.

Yet Another Turn

If the creationists were losing out in court, however, they were finding increasing success on a different field of battle. As a very vocal group, they have been able, by infiltrating local school boards and political groups, to make life difficult both for science teachers and for the publishers of school biology texts. Both of these groups, under constant harassment, eventually are tempted to sidestep a potential problem by downplaying or even ignoring evolutionary science, which is just what the creationists want.

Their fallback offer—that of presenting both sides—sounds like a reasonable compromise but really isn't. The cards, surprisingly, are stacked against evolution, just as they were against Huxley a century earlier when he "disappointed & displeased" everybody by trying to present all sides of the picture. The creationist approach is obviously simpler and, in many ways, more seductive, particularly when creationists convince unsophisticated congregants that accepting evolutionary theory means giving up Christ.

For the same reason, trying to debate evolution with a creationist in a formal setting is often a lost cause. Die-hard creationists may be "off the wall" as far as mainstream science is concerned, but that does not mean that they are stupid. Expanding on the idea of creation science, creationists have come up with yet another turn. Rather than just use the term "science" in their name, they have begun to argue "science" with the evolutionists. Certainly there have been many advances, many discoveries, many new facts to deal with. As a result, the complexity of the dispute has soared into the stratosphere.[34]

Further, scientific advances don't come easily, and there are plenty of discussions, and even arguments, among evolutionists.[35] The evolutionists say, "We are arguing science, not creed. That means getting into complex details." Too often, it also means the rapid loss of the public's ability to follow. The antievolutionary forces point to such arguments and compare these to their own much more settled point of view. "What kind of science is this," they ask, "where the scientists argue with one another?"

Equally interesting, however, is how the vast store of knowledge being built up in the biological sciences since Darwin's time is put to use by both sides: The same evidence is seen and used far differently by the two sides. Ronald Pine, a high school teacher and creationist watcher in Aurora, Illinois, puts it a bit more directly: A creationist "can come up with more lies in a half-hour than a scientist can refute in a week."[36]

Soapy Slides and the Problem of Complexity

The creationist "lies" can take many forms, including what could be called "soapy slides" around the basic argument. A common one maintains that systems in living things, such as the eye, are so "irreducibly complex" and so interdependent with all others, that it is just impossible to believe that all the systems could have come together via natural selection and resulted in a finely tuned, functioning whole. Somewhere there has to be some kind of "intelligent design." Richard Dawkins, in his attempt to come to grips with such creationist arguments, devotes a full 59 pages to the eye question alone.[37]

The so-called gap problem is another interesting example and shows up clearly in a long, lyrical, footnote-festooned article in the respected mainstream publication *Commentary*.[38] The author, David Berlinski, is not a biological scientist but does have some good credentials: He has taught mathematics and philosophy at the university level and has written a respectable book on the history of the calculus.[39]

The "gap problem" refers to gaps in the fossil record. There is no question that such gaps exist. A big gap appears at the beginning of the *Cambrian explosion*, over 500 million years ago, when great numbers of new species suddenly seem to appear in the fossil record. Unhappily, paleontologists have found relatively little connecting evidence to anything before that time.

The Cambrian example is perhaps the biggest gap, but there are many others, which is no surprise; much can happen to fossils in half a billion years. More important, however, is that many of the gaps are slowly being filled in, as new techniques—and more time—bring up new fossil evidence.

Referring to one of these major gaps, which lay between today's marine mammals and their (proposed) terrestrial ancestors, Stephen Jay Gould, a major player in the proevolution camp, writes, "I am absolutely delighted to report that our usually recalcitrant fossil record has come through in exemplary fashion [with] the sweetest series of transitional fossils an evolutionist can ever hope to find."[40]

Berlinski, in referring to a list of *250* gaps that have been filled, takes a rather different position: "That there are places where the gaps are filled is interesting, but irrelevant. It is the gaps that are crucial."[41] What he's saying is that as long as there are gaps, the theory can't be right. It's going to take a long time before he's satisfied.

Darwin was well aware of these gaps and devoted an entire chapter to them. He wrote, "I look at the geological record as a history of the world imperfectly kept, and written in a changing dialect; of this history we possess the last volume alone."[42]

Not surprisingly, Berlinski's article elicited a flood of responses; three months later, a follow-up piece, comprising many scholarly responses, was considerably longer than the initial article. Many comments were from evolutionists arguing against the author's points. One, from Daniel C. Dennett, was direct: "I love it: another hilarious demonstration that you can publish bullshit at will—just so long as you say what an editorial board wants to hear in a style it favors."[43] The journal also gave Berlinski another 16 pages in which to answer the charges and to continue his spiel.

Another soapy slide: Creationism is a descendant of the "argument from design," put forth almost 200 years ago by the British theologian William Paley. If you find a watch on the ground, what is the likelihood it has been put together by chance? Not much, you must agree. By the same token, what is the likelihood that a human has arisen in this way? Modern intelligent design (ID) theory equates evolution with chance and argues that intricacy must arise from design. With an eye on the courts, ID defenders leave the designer unnamed.

This reference to chance is highly annoying to evolutionists. According to Dawkins, many ID theorists who have trouble accepting Darwinian evolution are simply missing (or ignoring) an important point, which is that "Darwinism is not a theory of random chance. It is a theory of random mutation plus *non-random* cumulative natural selection"[44] (emphasis in original). "Why," he wonders, "is it so hard for even sophisticated scientists to grasp this simple point?"

Darwin faced a similar problem with Lord Kelvin, who, as a physicist, dismissed Darwin's biological evidence. Another Darwin contemporary, astrophysicist Sir John Herschel, called Darwinian evolution a theory of higgledy-piggledy.[45] "To this day, says Dawkins, "and in quarters where they should know better, Darwinism is widely regarded as a theory of 'chance.'"[46]

During the 1990s, yet another soapy slide has emerged in the evolution of creationism. Now that *design* has come to be recognized as a synonym for creation, the battle commanders offer still another new name for their pseudoscience: the "initial complexity model." If creationists have their way, this model will be taught alongside the "initial primitiveness model," their new name for evolution.[47]

At the same time, the conservative Christian view is still being pushed. An article in *Christianity Today* states that "the latest scientific findings support design over Darwin." Then it goes on to point out that "'Where do we come from?' is not an esoteric question relevant only to scientists. It is the beginning and basis of all we believe. And that's why Christians must come together, craft a credible apologetic, and then refuse to back down."[48] On the other hand, Catholic schools have long taught that the theory of evolution need not conflict with church dogma.

And so it goes. Somehow, the creationist army, in spite of all odds against it, seems to be growing. Ronald Numbers, an important creationism watcher, reports that creationists are now going after school boards; 2,200 out of 16,000 school boards in the country were "captured" by creation-leaning conservatives in 1992.[49] Though strongest in the United States, the movement is gaining adherents in other countries also.

Adoption of creationist doctrine can have serious results. According to one evaluation, it requires, "at a minimum, the abandonment of essentially all of modern astronomy, much of modern physics, and most of the earth sciences."[50] That statement was published in 1981. Since then, there has been an explosion of interest in and attempts to apply evolutionary principles to such fields as medicine,[51] pest control,[52] agriculture, and even psychology[53] and psychiatry, anthropology, ethics,[54] and sociology (e.g., the origins of behavior)[55]—to name just a few. New work in molecular biology is entwined here as well.[56]

A quick check of an online magazine database (UMI Research 1, including about 1,000 periodicals) for just the year 1996 brought up 1,349 entries for the word "evolution." This set of articles covers all aspects, of course: pro, con, and new work in the field (and probably some other, unrelated uses of the term). Even so, the number speaks for itself. (The entry *Bible* brought up fewer entries: 1,105.)

Wayne Grady, who reviewed one of Stephen Jay Gould's books for *Canadian Geographic,* has called evolution by natural selection "a theory that pervades every niche and cranny of our lives." He adds: "it ought, by now, to be law: the fact that it isn't, Gould reminds us, is a measure of our own inability to grasp so comprehensive a view of the world, not the result of any flaw in Darwin's general scheme."[57]

On the other hand, let us suppose that creationists got what they wanted. With the teaching that would result, the public's ability to evaluate scientific principles would surely weaken. It would become ever easier for nonsensical thinking to take hold.

Are we already on the way to such a situation? A 1993 Gallup poll revealed that almost half of all Americans believe that God created humans within the past 10,000 years. *Parade Magazine* reports: "75 percent of Americans cannot pass a basic National Science Foundation science quiz that asks questions like whether . . . humans and dinosaurs lived at the same time."[58]

What is happening appears to be part of an increase in fundamentalist beliefs, along with a broader rise in antiscience sentiments in general. And although the graduate science capability in the United States remains the finest in the world, science education in the lower grades seems worse than ever. This discrepancy cannot bode well for the future of this nation.

The introduction to this book mentioned History of Science Professor William Provine's belief that the ongoing war between science and religion can better be attributed to Darwin than to Galileo. We begin to see why he should feel this way.

Ironically, one of the first major publications to discuss this war was published in 1874, just 15 years after Darwin's *Origin* first appeared. Titled *History of the Conflict between Religion and Science,* it was authored by none other than John William Draper, the featured speaker at the Oxford debate. As David N. Livingstone puts it: "Draper's conflict metaphor proved to be captivating, and it provoked a whole series of similar military reveries."[59]

How ironic that the military metaphor should be stuck onto a debate that circles around Darwin. Throughout his life, he was, of all men, one of the gentlest, shyest, and most generous. With a deep love for and connection with nature, ever interested in and delighted by a flower, a worm, or a coral, he was the least warlike squire anyone could imagine.

Did his life and work challenge much of Victorian society to look at its religion, its science, and its morals in a new way? Certainly. But if he was, and still is, reviled by many, his remarkable achievement was, happily, recognized in his own day. Upon his death on April 19, 1882, he was given the quintessential recognition of achievement: burial in Westminster Abbey, near Newton. Nonetheless, the controversy he engendered rattles on, as loudly as ever.

CHAPTER 6

Lord Kelvin versus Geologists and Biologists

_____

The Age of the Earth

If we look back over the history of science and technology for someone who had it all, the name William Thomson comes up high on the list. Successful scientist, teacher, engineer, and businessman, he was showered with honors. There is not a home, office, or mode of transportation that has not in some way been touched by this man's work. By the end of his long and productive life, he had amassed 70 patents and had published more than 600 papers.

Born in 1824, Thomson seemed destined from his earliest days for scientific stardom. Educated by his father, a professor of natural philosophy, he matriculated at Glasgow University at age 10. After completing his studies there, he moved on to Cambridge University and graduated in 1845, at age 21, with honors. By 22 he was a full professor of natural philosophy at Glasgow University, a highly respected position.

Disdaining the long-used method of trying to pour knowledge, especially of the scientific kind, into hapless students' heads, he introduced the idea of illustrating his lectures with demonstrations. Once, to illustrate a point, he brought in an old muzzle-loader rifle and shot it at a pendulum.

Some students, mainly weak ones, complained that he was not a good lecturer. For those who could keep up with him, however, each of his lectures must have been a challenging experience. Although Thomson carefully prepared his first lecture,

105

he never prepared another one. As one of his early biographers put it, "Always a great quest! . . . No lecture was satisfactory unless some new fact or principle was wrenched from it."[1] If he was lecturing in one area, say on the stresses and strains of materials, and a thunderstorm came up, out came the electrometers and off he went in a totally different direction.

One day, his students decided to play a practical joke on him. He had prepared a raw egg and a boiled one and was going to show the difference in how they behave when spun. The students surreptitiously boiled the raw egg. As he began the demonstration, however, he quickly saw what had happened: "Both boiled, gentlemen," he said, smiling.[2]

When he began teaching, neither in England nor in Scotland was there a university research laboratory such as one finds today in any academic institution that teaches science. Even the illustrious Cambridge was not strong in experimentation, and Thomson later sought out laboratory experience in other scientists' laboratories. At Glasgow, he set up what appears to have been the first real laboratory for student use.

Although he remained at the University of Glasgow for more than half a century, word of his remarkable abilities spread rapidly, and for a large part of his life, he was widely thought of as the leading physicist and electrical engineer in the world. He was president of the Royal Society of London for five consecutive terms.

The prestigious *Dictionary of Scientific Biography* states, "Together with Helmholtz in Germany, he had been the foremost figure in transforming—indeed in creating—the science of physics as it was known in 1900."[3] He even played a part in creating the metric system—the international system of units being used in most of the world today. (He called the English system of weights and measures—the one still used in the United States but not in Britain—"barbarous.")

His interest in measuring processes and devices may have saved his life. In one lecture demonstration, perhaps with the muzzle-loader rifle mentioned previously, confusion between the avoirdupois dram (about 1.8 grams) and the apothecary's dram (about 3.9 grams) caused a student to put into Thomson's muzzle loader twice as much powder as he should have, which could have blown Thomson's head off. Fortunately, Thomson's finicky attention to details led him to check the amount with the student prior to carrying out the demonstration.

In fact, accurate measurement was one of his principal interests. "Can you measure it?" he wrote. "Can you express it in figures? Can you make a model of it? If not, your theory is apt to be based more upon imagination than upon knowledge."[4] The term *applied science* can also claim him as a progenitor, and he came up with numerous inventions, including improvements in the mariner's compass, sounding gauges for marine navigation, tide predictors, and a variety of sensitive measuring devices.

One of these measuring devices made it possible for Thomson to guide the laying of a successful transatlantic undersea telegraph cable between England and the United States in 1866, after an earlier first attempt had failed. The government expressed its appreciation by elevating Thomson to the peerage; he became Lord Kelvin in 1892, the first British scientist to be so honored. That's why another of his accomplishments, the absolute temperature scale, which has proved extremely useful in low-temperature physics, is named the Kelvin scale.

He was, in other words, a colossus, a megalith. At scientific meetings, he dominated the proceedings. But Thomson's ever-growing authority had a peculiar effect on a controversy that simmered and bubbled for an unbelievable 60 years.

The Age of Earth

The point at issue was the age of Earth. A century earlier, there had been little discussion about the subject. It was clearly stated in the scriptures, many claimed, that Earth is some 6,000 years old. The best-known voice was that of the seventeenth-century Irish bishop James Ussher. Using a complex combination of biblical chronology (mainly, counting up the "begats"), historical accounts, and astronomical cycles, he refined earlier estimates and, in the mid-1650s, came up with 4004 B.C. as the date of creation. The figure was used for 200 years in later English editions of the Bible.

Much of science in Ussher's time went to support such ideas and, in fact, many of the groundbreaking naturalists were also clerics. A good example is William Whiston (1667–1752), English theologian, mathematician, and astronomer. He was one of the first to introduce experiments into his lectures in London, yet he also used his own understanding of science to calculate that the biblical Flood that Noah survived had started on Wednesday, November 28, in the same year that Ussher had specified. He (and his fellow clerics) made many other, similar, estimates.

Another part of the scriptural reading was that cataclysms and catastrophes, such as the Flood of Noah's time, were the prevailing mode by which Earth's features were formed. The effects of these catastrophes were believed to explain the tortured appearance of much of Earth. According to catastrophism, Earth is both young and unchanging (ignoring such minor shake-ups as volcanoes and earthquakes).

The problem was that some of the new observations and theories began to contradict these Bible-based ideas. Buffon, whom you met in Chapter 4, was possibly the first in that strongly Christian era to try easing the age back beyond 6,000 years (i.e., before ca. 4000 B.C.). By estimating the cooling of Earth from an assumed former molten mass, he came up with an estimate of 75,000 years. More important than the number, which he later upped considerably, and more important even than the contradictory nature of these results, was the implication that nature was rational and would give up its secrets to those who learned to read and understand its language.

Another early worker in the search for the age of Earth was Benoit de Maillet (1656–1738), of France. An amateur naturalist, he did his calculations based on an observed decline of sea level. Interestingly, he came up with a figure of 2 billion years, which comes much closer to the modern figure.

To help protect himself from reprisals, de Maillet presented his findings in a story based on a series of fictitious conversations between a French missionary and an Indian philosopher named Telliamed (de Maillet spelled backward). Perhaps recalling Galileo's treatment, he nevertheless hesitated to publish;

his account did not appear in print until 1748, 10 years after his death, and had little effect on the situation.

There were other attempts to determine Earth's true age, and by Thomson's time, a plethora of estimates had been put forth, based on a wide variety of methods. The most authoritative and effective refutation of the Christian idea of a very young Earth, shaped by catastrophes, was that of the respected British geologist, Sir Charles Lyell (1797–1875). Lyell argued that catastrophism was not necessary and that Earth's features could be explained by forces still working. In fact, he believed, everything seen on Earth is a result of ordinary forces and agents, all of which act in a uniform manner. (His theory was therefore called "uniformitarianism.") Uniformitarianism meant that the past could be explained in terms of what we see happening today.

From a contemporary viewpoint, the major importance of uniformitarian theory is that there was no need for such catastrophes as the Flood or for any other supernatural influences. If Lyell was right, a literal reading of the Bible was no longer a tenable route for science. His doctrine also, however, called for these forces to have been acting over unlimited time!

By the mid–nineteenth century, uniformitarianism (a label, ironically, coined by catastrophist William Whewell) had become the dominant geological doctrine in England. Although the theologians were not happy with the uniformitarian doctrine, stability is for most of us a more comforting state of affairs than the idea that we could all be wiped out at any moment. In Chapter 3, we saw Newton and Leibniz arguing about God's place in the stability of the solar system. Early in the nineteenth century, the French mathematician Laplace finally showed that God need not act as an after-the-fact watchmaker after all, that the system was quite stable on its own. Many people heaved a deep sigh of relief.

While Laplace thought this stability was just a chance arrangement, a matter of good luck, many others disagreed, feeling that it showed clear evidence of God's hand. Thomson was one of them. At the same time, however, observations on Encke's comet appeared to show that there is some sort of resisting medium in interplanetary space. This suggested to Thomson the eventual running down of the entire system—and it fitted in well with other work in which he was deeply interested, and which brought together several aspects of his diverse interests.

Thomson's Line of Thought

Ever since his student days, Thomson had a warm place in his heart for the subject of heat. He was undoubtedly aware that Leibniz had earlier been an important believer in an initially molten Earth, and that Newton had done some work on heat loss and the cooling of bodies. By age 18, Thomson had already published a paper on the "Uniform Motion of Heat in Homogeneous Solid Bodies and Its Connection with the Mathematical Theory of Electricity." The title is significant, for it shows that he not only was interested in the problems of heat and its movement through solid bodies, but also was already trying to apply to the problem of heat motion the mathematical methods that had proved so successful in dealing with mechanical motions and electricity.

For a hands-on type of person, Thomson's grasp of mathematics was remarkable. For example, he followed the work of Joseph Fourier, who had done some pioneering mathematical work in heat conduction. Using the calculus of Leibniz and Newton, Fourier had derived a way of finding at any time the rate of variation of temperature from point to point in a solid, as well as the actual temperature at any point in that solid. Thomson was fascinated by the method. He wrote later that, although still a student at the time, "In a fortnight I had mastered it—gone right through it."[5]

Although he later referred to Fourier's work as a "great mathematical poem,"[6] it served a more plebeian purpose, for it helped convince Thomson that Earth had undergone a continual cooling from an initially hot, fluid state to its present condition.

Earlier, the French physicist Sadi Carnot, influenced by the enormous importance of the steam engine, had shown that heat and work are interconvertible. But little attention was paid to this important idea until Thomson, in 1849, looked more closely and came up with an important advance.

Thomson was convinced that some portion of the heat input was not available for work, which was an important fact to know when designing such machines. In addition, however, he broadened the focus to include the part played by these phenomena in the workings of Earth.

To his mind, a tantalizing clue to the age of Earth lay in a common observation made in the digging of countless mines and wells: The farther down you dig, the hotter Earth gets. Although there are other possibilities to explain this phenomenon, Thomson believed that it showed that heat was flowing out from Earth's interior.

As he understood it, heat energy is escaping from Earth and, as in the steam engine, is basically not recoverable. This dissipation of energy implied a running down of our natural systems and became, in a paper he presented in 1851, the second law of thermodynamics, one of the rock-solid bases for the scientific application of heat and work. The first and second laws, roughly stated: None of the energy is ever lost (first law), but some portion is not available for work either (second law).

The second law provided a quantum leap in science's understanding of physical machines of all kinds. It showed finally, for example, why perpetual-motion machines are impossible. It also tells us, said Thomson, that natural engines—such as the Sun, Earth, and other parts of the solar system—must run down as well.

In his calculations, he started with the assumption that Earth was initially part of the Sun, had originally been at the Sun's temperature, and that it has been cooling off continually and steadily ever since. At first, Thomson used his calculus in an attempt to get a feeling for how long Earth and the solar system might hang around in roughly their present state. Then, in an 1842 paper, he considered the possibility of working the calculations *backward* instead of forward. Suddenly, it appeared possible to calculate the age of Earth with some sort of scientific accuracy.

Recognizing some weaknesses in the approach, he began to refine it and developed the ideas further in the following years.

In 1846, the same year of his appointment at the University of Glasgow, he reported his calculation of the age of Earth based on physical principles. Everyone sat up and listened. The time required for Earth to reach today's temperatures, he stated, was about 100 million years. Recognizing that the figure was in truth an approximation, due to his simplifying assumptions, he widened his net to somewhere between 20 million and 400 million years.

Debate

If Thomson was right, however, then several major theories were unworkable. The geologists, for example, looked about them and saw a tortured Earth that cried out for a history stretching back billions of years. Darwin's theory of evolution, still struggling to take hold, also required a much longer prehistory than Thomson's figures allowed. As a result, Thomson never accepted evolutionary theory.

In our own time, Thomson has been advanced by the creationists as an example of a major scientist who believed in their creed. This, however, is a major misuse of scientific history. Although Thomson rejected Darwin's evolution, he was in no way a creationist; that is, he did not align himself with religious literalists, and his objections were in no way similar to the religious attacks on evolution that continue to plague the world of biological science.

Even though Thomson was stamping on the ideas of many mainstream scientists, he never felt that he was standing alone. James Prescott Joule, who had done solid work in demonstrating the mechanical equivalent of heat, was one of his supporters. In a letter to Thomson dated May 1861, Joule wrote, "I am glad you feel disposed to expose some of the rubbish which has been thrust on the public lately. Not that Darwin is so much to blame because I believe he had no intention of publishing any finished theory but rather [that he wanted] to indicate difficulties to be solved. . . . It appears that nowadays the public care for nothing unless it be of a startling nature. Nothing pleases them more than . . . philosophers who find a link between mankind and the monkey or gorilla."[7]

By 1869, Thomson had aligned himself with what he called the "true geologists," meaning of course those who agreed with his time scale. As for those other geologists, and the biologists, they needed help. Therefore, nine years after Thomas Henry Huxley's famous debate with Bishop Wilberforce, Huxley again found himself acting as a public advocate. Though remembered today as Darwin's bulldog, Huxley was a distinguished scientist in his own right and had served as president of the Royal Society of London, which is why he was chosen to do battle with Thomson.

This time, however, the debate was held in a more scientific arena—the Geological Society of London. There was another important difference: Huxley was crossing swords with a far more capable adversary—Thomson, who, incidentally, had attended the earlier Wilberforce–Huxley debate. (It should be noted that the verbal debate between Huxley and Thomson in no way settled anything. It carried over into writing in the following years, and it drew in many other entries as well. We'll draw from all of these sources in this chapter.)

Thomson's understanding of Darwin's work, and Huxley's arguments in the debate, led them into some rather deep waters—namely, the origins of life on Earth. Huxley's approach can be summarized in his 1870 Presidential Address to the British Association for the Advancement of Science (BAAS), in which he stated, "if it were given to me to look beyond the abyss of geologically recorded time to the still more remote period when the earth was passing through physical and chemical conditions, which it can no more see again than a man can recall his infancy, I should expect to be a witness of the evolution of living protoplasm from not living matter."[8]

Thomson jumped on this aspect and used it in his dismissal of evolutionary theory; he maintained that science had given us "a vast mass of inductive evidence against this hypothesis of spontaneous generation."[9] This was somewhat unfair, in that there is far more to evolutionary theory than the earliest beginnings. Nevertheless, Huxley's approach to the origins of life was a remarkable, and fair, statement that could still stand today.

Thomson would have none of it, however, and insisted that life had to come from life. His explanation sounds at first like the more scientific one: "If a probable solution, consistent with

the ordinary course of nature, can be found, we must not in-
voke an abnormal act of [creative power]."[10] The only other
route he could imagine was that "there are countless seed-
bearing meteoric stones moving about through space," and that
some of these, landing on Earth, provided the necessary beginnings of life.[11]

Huxley, in a letter to a colleague dated August 23, 1871, responded: "I like what I see of Thomson much. He is, mentally,
like the scene which lies before my windows, grand and massive but much encumbered with mist—which adds to his pic-
turesqeness but not his intelligibility."[12] Huxley also asked another colleague, Joseph Dalton Hooker (a friend of Darwin),
"What do you think of Thomson's creation. . . . God almighty
sitting like an idle boy at the sea side and shying aerolites (with
germs), mostly missing, but sometimes hitting a planet!"[13]

Yet another dart at Thomson came in the form of a bit of
doggerel in a local publication:

From world to world
The seeds were whirled
Whence sprang the British Ass[14]

(Ass being an irreverent nickname for the British Association
for the Advancement of Science, in which both Thomson and
Huxley played an active part).

Of course, Thomson's statement about life-bearing meteorites only pushes the problem backward; in truth, we are not
much further along today in our understanding of the matter.
Still, it is delightful to read current scientific reports that a
Stanford University team has found what may be relics of ancient life in a Martian meteorite that landed on Earth.[15]

At the time of the debates, however, Darwin and his cohorts
were still squeezed by Thomson's work. One adjustment they
tried was to shorten the time needed by evolution to do its
work. George Darwin, one of Charles's sons, who had grown
into a respected scientist in his own right—and had earlier
worked with Thomson!—tried to defend his father. In a letter
to Thomson dated 1878, he wrote, "I fail to see the justice of
your remark that a few hundred million years would be insuffi-

cient to allow of transmutation of species by nat[ural] selection. What possible datum can one have for the rate at which it has or can work?"[16]

Although Thomson's opponents all accepted the accuracy of his calculations, some felt that there was a different problem that was not being adequately addressed—too many assumptions and not enough solid scientific data. Huxley later wrote, "mathematics may be compared to a mill of exquisite workmanship, which grinds you stuff of any degree of fineness; but nevertheless, what you get out depends on what you put in; and as the grandest mill in the world will not extract wheat-flour from peascods, so pages of formulae will not get a definite result out of loose data."[17] Also, "this seems to be one of the many cases in which the admitted accuracy of mathematical processes is allowed to throw [on the subject] a wholly inadmissible appearance of authority."[18]

Another critic, Fleeming Jenkin, suggested that one of Thomson's calculations "savors a good deal of that known among engineers as 'guess at the half and multiply by two.'"[19] Their quite valid objections had little effect, however. Unfortunately, they had missed Thomson's basic point, which was that if *any* limit whatever could be placed on Earth's age, it would refute uniformitarianism. Thomson felt strongly that as long as the geologists supported uniformitarianism, geology would remain an inexact science, one that depended on hypothesis and guesswork.

As for the debate itself, it had, like the earlier one between Huxley and Wilberforce, brought the age-of-Earth question into the public forum and had generated great public interest. The result, however, was that both scientific and public support moved only further in Thomson's direction.

In 1894—two years after Thomson's elevation to the peerage as Lord Kelvin—Lord Salisbury, president of the BAAS, still maintained that Kelvin's figures remained one of "the strongest objections" to Darwinian evolution. The geologists and biologists, he felt, had "lavished their millions of years with the open hand of a prodigal heir indemnifying himself by present extravagance for the enforced self-denial of his youth."[20]

Even Mark Twain got into the act. Some time around the turn of the century, in a brief sketch called "Was the World

Made for Man?" he wrote, "Some of the great scientists, carefully ciphering the evidences furnished by geology, have arrived at the conviction that our world is prodigiously old, and they may be right, but Lord Kelvin . . . feels sure it is not so old as they think. As Lord Kelvin is the highest authority on science now living, I think we must yield to him and accept his view."[21]

Although the frustrations of Thomson's opponents may well be imagined, and in spite of what may sound now like some pretty strong language, the disputants somehow managed to coexist and to maintain reasonably good relations, right up to the end of the century.

Still, as the century waned, something was happening. Even Kelvin himself, as he was now called, began to wonder whether he had been too limited in his view. By 1894, he was thinking that perhaps 4 billion years might have been a more appropriate upper limit for the age of Earth. The prevailing view of Kelvin as utterly inflexible may be too harsh. By then, however, it no longer mattered, for the original figures had hardened into stone. His calculations had been used as classical examples for 30 years by students in physics classes the world over.

We know today that the geologists and the biologists were right in their claims for a much more ancient Earth than Kelvin had originally calculated. In a strange bit of irony, however, it took entirely new methods, developed by physicists, to supply the needed proofs that he was wrong. What Kelvin didn't know, and what no one in his day could know, is that there is indeed an additional major input of heat within Earth.

New Findings

The beginning of the end for Kelvin's calculations came with the discovery of radioactivity by the French physicist Antoine Henri Becquerel in 1896. Although it took some further years before the process became clear, Pierre Curie and Albert Laborde showed in 1903 that, thanks to this radioactivity, radium had the amazing ability to radiate heat continuously. As a result, the material did not cool down to the temperature of its cooler surroundings, which is the way most warm objects behave.

It was further found that the various radioactive elements were not independent elements but could, in some way, be descended from one another. Radium, for instance, derives from uranium, and lead is the final stable product of uranium disintegration. In 1907, Bertram Borden Boltwood, an American chemist and physicist, suggested that because we know the rate of disintegration of uranium ore into lead, if we determine the amount of lead in a specific uranium ore sample, then it might be possible to determine the age of the rocks in which the ore is found. The higher the percentage of lead in the ore, the older it is.

Further development of such methods has led to today's far more accurate set of dates. The earliest date for a rock sample found on Earth is about 4.3 billion years. It seems fair to assume that Earth is older than the oldest rock samples. How much older? Evidence from meteorites suggest an age for the formation of the solar system of about 4.6 billion years. Recent work using other equipment, such as lasers and ion probes, has so far borne out the earlier calculations.

Other discoveries have brought to light facts about Earth that could only be guessed at in Kelvin's day. We know, for instance, that heat from several causes—gravitational energy and meteoritic bombardment, as well as internal radioactivity—has caused and maintained a partial melting within Earth. The result is a powerful convectional process, a mixing and uplifting of molten rock, as well as the conduction of heat from inside to outside that Kelvin worked with. There has also been a chemical segregation that he couldn't have known about. The final result of all these processes, which still go on today, is a thin crust of outer rock, a rock mantle of greater density, and a still heavier core of iron and nickel.

Further, the powerful convection processes exert enormous forces on the various parts of our planet, which fold, bury, tear apart, and lift up great areas of Earth's surface. The result is that none of the earliest rocks is likely to have survived, which is why even the oldest of dated rocks can't tell the whole story.

Now new work has surfaced suggesting that there are yet additional major forces at work, which may skew scientific results even further. Debra S. Stakes, a geochemist at the Monterey

Bay Aquarium Research Institute, argues that "most geological processes at their more fundamental stages could be biologically mediated, which challenges our models for inorganic thermodynamics, for driving reactions. Microbes have been found thriving more than two and a half miles down, at a temperature of 230 degrees F. The cumulative mass of these organisms may exceed that of all the inorganic matter that makes up our earth. More and more, it appears that microbes, dwelling miles deep in the earth's crust, have played a major role, perhaps the dominant one, in creating and arranging the rocks, soils, metals and minerals, as well as the seas and gases."[22] In other words, we still have plenty to learn about Earth's origins and activities.

Denouement

Wrong as Kelvin was, however, his reputation never waned. Even though the new dating methods showed clearly that Kelvin's figures were wrong, and even though he refused to accept the reality of radioactivity, he remained a powerful and respected figure in science. In 1904, at the age of 80, he became Chancellor of the University of Glasgow.

Lord Kelvin's prominence put Sir Ernest Rutherford in a very ticklish position when he was invited in the same year to address a meeting at the Royal Institution. Rutherford had been doing highly important work in understanding what was going on in the atom, and he had already shown that radioactive atoms, and perhaps all atoms, contained great stores of latent energy. He knew well that this new information brought him into sharp conflict with Kelvin, who was in the audience, regarding the age of Earth.

"To my relief," Rutherford later wrote, "Kelvin fell fast asleep, but as I came to the important point, I saw the old bird sit up, open an eye and cock a baleful glance at me! Then a sudden inspiration came, and I said, 'Lord Kelvin had limited the age of the earth, *provided no new source of heat was discovered.* That prophetic utterance refers to what we are now considering tonight, radium!' Behold! The old boy beamed upon me."

The old boy may have beamed upon Rutherford; he may even, as noted earlier, have had a few doubts about his age-of-the-Earth calculations; but that doesn't mean he had actually changed his mind. As late as 1906, he was still maintaining that radioactivity could not account for the heat of Earth. Many geologists were equally unable to adjust their thinking. It took another decade before a new generation of scientists were able to free themselves of Kelvin's influence.

Although Kelvin's timing was off, his basic point—that the solar system is running down—remains fundamentally correct. Fortunately, it will take a lot longer for the end to come than he thought it would.

At Kelvin's death in 1907, he was buried in Westminster Abbey, next to Newton and near his old foe, Darwin.

CHAPTER 7

Cope versus Marsh

The Fossil Feud

On the morning of January 12, 1890, the scientific community received the shock of its life. A feud that many of its members had known about for years was suddenly splashed across the front page of the *Herald*, a major New York City newspaper. The bold headline read

SCIENTISTS WAGE BITTER WARFARE

Following were nine columns of juicy detail, in which Edward Drinker Cope, of the University of Pennsylvania, advanced serious charges against Othniel Charles Marsh, who was not only Professor of Paleontology at Yale University, but also president of the National Academy of Sciences and an important member of the United States Geological Survey. The charges included, but were not limited to, plagiarism, incompetence, and even the smashing of fossils to prevent others from getting at them.

Suddenly, a feud that had simmered and boiled for two decades was opened to the public, with a variety of consequences:

- The *Herald* sold lots of newspapers; subsequent issues over the next two weeks continued to carry charges and countercharges. Marsh countered by accusing Cope of, among other things, stealing some of his fossils, sneaking into his private workrooms, and even being mentally unbalanced. Both Cope and Marsh, clearly, had been building a detailed

file of damaging information about each other for many
years, and each provided plenty of information to the
Herald.

- A few scientists, including Marsh himself, claimed that
they were happy the whole thing was finally being brought
out into the open.

- The vast majority of scientists, however, particularly those
in the worlds of paleontology and geology, were embarrassed, to say the least. Most scientists approached by the
Herald reporter who was working on the series simply refused to get involved.

Although Cope and Marsh had never come to blows physically during the course of their feud, each used just about every
other means to bring down the other, including disputing each
other's priority and conclusions on every possible occasion.
Also, though both aspired to be the premier fossil collector and
expert in the country, it seemed almost that their race had
more to do with spiting the other than with advancing the science of vertebrate paleontology. There should have been plenty
of room for both of them in the wide American West, where
most of the excavations took place. As you'll see, however, this
was not the case.

The Setting

Cope and Marsh were not the first, by any means, to unearth
dinosaur fossils. Scattered finds of dinosaur-bone fragments in
Europe, starting in the late 1820s, led researchers to a realization that something interesting was afoot. Richard Owen, a
British comparative anatomist and early paleontologist (mentioned as a foe of Huxley and Darwin in Chapter 5), finally proposed in 1842 that many large reptilian bones found in southern England were from a now-extinct group of reptiles. He
suggested the name *dinosaur,* from the Greek *deinos,* terrible,
and *sauros,* lizard. He was a little off in details but did recognize these creatures as large land-living reptiles that differed
from any known living reptiles.

A contemporary description in an 1855 literary journal referred to some restorations created by Owen: "But Heaven help us," the author cried out, "what are these?—these frightful scaly monsters—these giant reptiles—these gaping jaws, and eyes in which no speculation dwells?"[1] Dinosaurs, of course, did not have scales.

By then, evolutionary ideas were already in the air. But when Darwin finally published his *Origin of Species* in 1859, the fat was in the fire; his supporters began to look toward the fossil record for support. Eight years later, Thomas Henry Huxley published a paper pointing out a strong similarity between certain of the extinct dinosaurs and our birds of today. Although the idea didn't gain much support then, it has again come to the fore and has led some researchers to argue that dinosaurs are not extinct but have evolved into and live on today in our bird population.[2] A happy thought.

Others managed to see the opposite. Owen, for example, chose to believe that dinosaurs disproved evolutionary theory. Also, at the time, the fossil record did indeed seem to show that new groups of animals arose de novo, did not evolve from some earlier strain, and died out without leaving any more-advanced descendants. He remained a staunch antievolutionist all his life.

On this side of the Atlantic, our two combatants also took opposing sides, with Marsh for evolution and Cope against it. But both men were to have a powerful effect on the outcome of the argument. In 1865, just two years before Huxley's paper was published, the Civil War had ended, and Americans were ready to turn to other matters. These included pushing the transcontinental railroad across the wild reaches of the Midwest. The vast amount of blasting and digging began to unearth a diverse sampling of strange-looking bones.

Little was understood about the early history of any animal life, however, let alone dinosaurs. These creatures, we know today, arose during the late Triassic period, more than 200 million years ago. They lived and flourished for an astonishing 140 million years, finally dying out about 65 million years ago.

But a lot can happen to Earth's surface in 65 million years, so it's no surprise that people do not commonly find dinosaur bones. Yet there were places where broken pieces of dinosaur bones were actually lying about. The problem is that if you don't know that something (e.g., a dinosaur) exists, it's hard to tell whether what you're seeing has any significance. The first known finder of such specimens, a sheepherder, used the dinosaur fragments to construct a cabin.

This activity took place in an area named Como Bluff, a long east–west ridge located in southern Wyoming, which turned out to be one of the world's great storehouses of dinosaur specimens. There were other such areas, however, and bone fragments, as well as complete pieces, were being dug up in a variety of places.

One of these sites, surprisingly, was Haddonfield, New Jersey, where, in 1858, one of the first reasonably complete dinosaur skeletons had been found. It was identified and described by Joseph Leidy, a leading American paleontologist. A decade and a half later, Leidy had his own kind of run-in with Marsh and Cope. What sort of men were these two?

Marsh

Born into a farm family in Lockport, New York, in 1831, Marsh's mother died when he was only two years old. Although his father remarried, his early personal life seems to have been difficult. But he developed a strong interest in fishing and shooting, and the open-air life gave him the vigorous health he enjoyed for many years. He also developed an interest in fossils, which were frequently exposed by the digging operations for the Erie Canal being widened nearby.

In 1852, Marsh's rich uncle, merchant and philanthropist George Peabody, heard about Marsh's interests and began to fund his education. Because of Marsh's late start, he was older than his college classmates and, although not disliked in school, he was not considered a good mixer. The daughter of the people with whom he roomed wrote afterward: "Mother says he was always very odd and for most people it was 'like running against a pitchfork to get acquainted with him.'"[3]

The outlines of his future in collecting were already showing, however. He wrote in a copybook, "Never part with a good mineral until you have a better."[4] He spent his summers on expeditions, collecting minerals and fossils. He lived well and began a collection in his own rooms. His first scientific paper, on a gold field in Nova Scotia, was published in 1861—at age 30, and while he was still in school. He graduated Phi Beta Kappa from Sheffield Scientific School (part of Yale) in 1862.

Marsh's performance impressed Peabody, who provided substantial support for a variety of good causes. Marsh's career became one of them. After graduating from Yale, Marsh went to study in Europe, a common route for budding American scientists. He also visited Peabody, who was living in London, and persuaded him to contribute substantial funds for a new museum at Yale; it eventually became the world-renowned Yale Peabody Museum, and Marsh's home base.

Marsh returned from Europe in 1865 and, still with Peabody's financial support, was able to take an unpaid position as Professor of Paleontology at Yale. This turned out to be a smart move. The Yale connection was invaluable, and because he had no teaching requirements, he was free to pursue his love of collecting and research.

Although most of his early expeditions were in the eastern part of the country, he began to hear about fossil finds in the Midwest. He made his first foray into that still wild area in 1868, the first of about a dozen expeditions to a variety of regions along the eastern flank of the Rockies, of which the earlier ones were financed entirely by his own—that is, Peabody's—funds.

These expeditions were also attended by considerable danger and hardship, and his early training in the out-of-doors stood him in good stead. Using some of the connections he had developed, he was able to arrange for military protection when moving into American Indian territory. This included the assistance of the famed Buffalo Bill Cody, who acted also as a scout. In many cases, his protectors also doubled as fossil hunters.

Fossil hunting, even then, was not a matter of going for a hike and keeping an eye open for interesting things sticking up in the air. Choosing the right region meant knowing at least approximately what geological era was exposed and why. Was a particular specimen really old and from that area and era, or was it brought in by a later flood or other natural occurrence? Was the fossil hunter working in a flat area, or was the exposed surface tilted by an uplift of some kind? The digging techniques for each of these situations differ greatly.

Marsh's assistants, both in the field and back in New Haven, were to be of great importance in his career—and also a source of constant frustration. Several of them turned on him when the occasion arose, and they sided with Cope in the newspaper war. One reason for their betrayal was his cavalier treatment of them. In some cases, he either held back on their salaries or simply let the matter of payment slide, sometimes for a month or more. When sending out new workers into the field, he tended not to set up lines of command. It's hard to say whether this tendency resulted from his own freelance mentality; from administrative weakness (unlikely); or at least partly from a desire to deliberately stimulate competition among his hired hands.

The results ranged from successful to near catastrophic. One confrontation actually brought a gun into play. Fortunately, the gunless antagonist, pointing out that he had a family, backed off.

One of Marsh's workers, William Harlow Reed, discovered a new quarry about a mile from the railroad. He requested money to hire a horse to carry specimens from the site to the station. Marsh never even bothered to reply, so Reed had to carry heavy loads on his back across a dangerously swollen creek.

Marsh also insisted that all publications concerning the fossil finds were to be published under his name alone!

Why did his workers stay with him? Partly, it was a job, which was not always easy to get in those days. There must have been additional reasons, however. Possibly there was the excitement of being in on the initial stages of this work. While working in the Como area, Reed once wrote back to Samuel Wendell Williston, another of Marsh's assistants, "I wish you wer hear [sic] to see the bones roll out and they are beauties . . . it would astonish you to see the holes we have dug."[5]

One explanation for Marsh's self-centered approach to life was that he never married, so he never learned the art of sharing. George J. Brush, director of the Sheffield Scientific School, supposed that Marsh remained a bachelor because he would never be content with anything less than a collection of wives.[6]

Considering all of Marsh's activity, his zeal to be first and foremost, and the primitive state of knowledge in that field, it's not surprising that he, like Cope, made some mistakes. In one case, his collectors had come up with an almost complete skeleton for an unusually large beast. Unfortunately, the all-important skull was missing. His first mistake was to jump to the conclusion that he had in hand a new species, which he named *Brontosaurus* (thunder lizard). The skeleton was later assigned to an existing species, *Apatosaurus*. Although *Apatosaurus* is now the proper name, Marsh's term is still commonly used and is a source of constant confusion.

Much worse, however, was the way he solved the problem of the missing skull. Because no one knew what it should look like, and because Marsh had plenty of other skulls and parts of skulls lying about, in his anxiety to reach closure on the project, he simply assigned the skeleton a skull from an entirely different species. The result was that for a hundred years, this major specimen was displayed with the wrong head—and so was every other specimen elsewhere that was based on his reconstruction. Early in our own century, other researchers suspected a problem; as was the case with Kelvin, so influential was Marsh's reputation that it was not until 1979 that the mistake was fully rectified.

On the other hand, although self-centered and anxious to build his credits at anyone's expense, he could also be generous and helpful. In the mid-1870s, he became involved with a group of Sioux Indians. In return for being permitted to pass through their land, he promised to plead their cause to the government's Bureau of Indian Affairs. To the tribe's surprise, he carried through and helped them attain certain of their objectives. Among the Sioux and, as his fame grew, other tribes as well, he became known as the Bone Medicine Man and the Big Bone Chief.

The fossils pouring into his headquarters at New Haven were beginning to pile up, however, and after the 1874 expedition, he made only occasional, brief trips out to the diggings to check on the progress of his various hired field parties.

Williston, who worked for Marsh from 1874 to 1885, wrote later, "The real cause of his inadequate productivity after 1882 was that he had become overwhelmed and confused by the very mass of his fossil riches, and by the effort required to direct his superabundant staff in the lab and in the field. At times, his assistants were left for a day or more with nothing to do except talk over their grievances, while Marsh lingered in New York at the University Club or the Century Club, where he undeniably liked to 'spread himself.'"[7]

Cope

Cope's early days had several similarities with Marsh's. Born in 1840 on an 8-acre farm near Philadelphia, he was only three years old when his mother died, while giving birth to her third child. His father remarried but appears to have continued being the main influence on him. Cope's early schooling and much of his home life were submerged in Quakerism. He didn't take to farming, but farm life opened up the world of nature to him; he collected animal and plant specimens and made careful notes on them.

After high school, Cope convinced his father to let him attend Leidy's program in comparative anatomy at the University of Pennsylvania. He also did some work at the herpetological collection of the Academy of Natural Sciences in Philadelphia. In 1863, he too went off to Europe, ostensibly to continue his studies. There is also the possibility that he was sent over by his father to escape the Civil War draft (Quakers vehemently opposed both slavery and war) or to get over some sort of connection with a girl. While there, however, Cope made good use of the museums and managed to meet many of Europe's distinguished naturalists.

On returning to the United States in 1864, he assumed management of a farm his father had bought for him, while also beginning a teaching career at Haverford College in Philadelphia. He married in 1865 and left both the farm and his post at Haverford, moving to the Haddonfield area of New Jersey around 1867, in order to be close to the fossil beds.

Using funds realized from a farm he had inherited from his father, Cope decided to become a freelance scientist. He remained in Haddonfield until 1876, honing his skills, then moved back to Philadelphia and bought two adjoining houses. For many years, he lived in one with his family and used the other as a personal museum and storehouse for the ever-increasing load of fossils he was to collect.

Cope worked hard and was extremely productive; although born nine years after Marsh, he started his career as a paleontologist at a much younger age. He published his first article at age 18, and by his 20s, he had already established an international reputation as a herpetologist (reptiles and amphibians) and ichthyologist (fish). He produced the first comprehensive account of North American snakes; also, *Copeia,* a leading journal of American herpetology and ichthyology, is named after him. Some of Cope's scientific reputation developed only after his death, when the dust from the feud settled a bit, and when the scientific world learned enough to know what he had been able to accomplish as a kind of lone ranger.

Cope's greatest pleasure came in describing new species, often using ingenious polysyllabic Greek names. Unfortunately, however, when his terms were handled by untutored telegraph operators and typesetters, who were being pressured by him to put the notice through in a great hurry, the spellings were sometimes less than accurate, and the results often came back to haunt him.

Cope had other problems as well. Marsh, a better politician, had found ways to get the government to help fund his expeditions, and he tried to turn some of the public collecting areas into his own private reserves. When Cope found himself falling behind in his competition with Marsh, he felt that he had to redouble his collecting efforts, which required him to hire more assistants. In 1881, in an effort to obtain the needed funds, he began to gamble on unfortunate mining ventures. By 1885, he had pretty much used up both his wealth and his robust health, and the realization that Marsh was winning the fossil-finding war ate at him.

Nevertheless, even during the mining years, which lasted until 1886, he redoubled his scientific output. Working with fossils that had already been collected, he published dozens of papers a year, some of them extensive reviews of major animal groups. Over the course of his career, he produced more than 1,400 scientific papers and monographs on fossils of all kinds, as well as on living creatures. An indication of how difficult this work was can be seen in the fact that more than a century later, one of the groups he named, the genus *Coelophysis,* remains the center of an ongoing controversy.[8]

To offset his financial problems, he had tried to sell his collection, but without success—until, with help from his friend Henry Fairfield Osborn, the American Museum of Natural History in New York finally purchased a significant part of it, in 1885. Though impressive, and a good start for the museum (which Osborn was later to head), it was far smaller than Marsh's collection, valued at over a million dollars even in those days.

Like many other brilliant people, Cope was complex. Well liked by some, he clashed early with others. Right from the beginning, he had had problems with officials, including run-ins with the administrators at Haverford College. He was reprimanded for not following orders while working on one of the geological surveys being carried out in the West. Later, he quarreled with council members at the Philadelphia Academy of Natural Sciences, where he still put in some time, and he finally resigned—or maybe was pushed out.

Terms used to describe him included friendly, considerate, altruistic, generous, honorable, manly, and devoted to his family; also, however, he was described as straightforward, ascetic, militantly independent, uncommonly candid, and sincere—all the latter being qualities that were sure to irritate some who came into contact with him.

Was he off balance, as Marsh claimed in the *Herald?* There were indications along this line. It's possible, for instance, that an intensive Quaker upbringing and an unusually intense supervision of his education by his father may have had a negative effect on his later life. He certainly went through some difficult times: He experienced some unexplained illnesses; at least part of the reason for his trip to Europe was an attempt to get over some sort of mental problem; he went through a difficult, introspective period while in Europe, including fears for his own sanity and some destruction of his notes and drawings; he was, at least for a time, a religious fanatic and required his diggers to listen to Bible readings after digging hours; and he had severe and horrifying nightmares on at least one expedition.

E. C. Case, a writer who had known Cope briefly, wrote much later (1940) in *Copeia,* "The clue to an understanding of Cope's life is the realization that he was essentially a fighting man, expressing his energy in encountering mental, rather than physical difficulties. . . . He met honest opposition with a vigor honoring his foe, but fraternized cordially after the battle."[9] That approach characterized even his relationship with Marsh—at first.

Conflict

Both Cope and Marsh were men of independent means. Edwin Colbert, a major figure in contemporary paleontology, suggests that as a result, "being free from the necessity of making those daily adjustments that are the lot of most men they both lacked a certain amount of perspective in the field of human affairs. They were both possessive and ambitious to an unusual degree." Further, Colbert says "Neither was overly scrupulous."[10]

It was, therefore, perhaps inevitable that there would be trouble. It didn't emerge right away, however. In fact, the two men appeared to be quite friendly at first. They visited back and forth, went out on a dig or two together, named species for each other, and corresponded in a friendly way.

What went wrong? When did friendship turn, or begin to turn, to enmity? The answer depends on whom you ask. In the *Herald* series, Cope claimed that in the early days of their relationship (1868), he had taken Marsh "through New Jersey and showed him the localities" where the first American dinosaurs from the Cretaceous period had been found and had been written up by Cope. He continued, however, "Soon after, in endeavoring to obtain fossils from these localities, I found everything closed to me and pledged to Marsh for money considerations."[11]

Another possibility is that the hard feelings trace back even earlier, to 1866, when Cope showed Marsh some work he had done on the skeleton of a plesiosaur found in Kansas. Marsh discovered a serious error in a drawing Cope had made of this aquatic reptile. To put it bluntly, the head was attached at the wrong end of the skeleton. Marsh published his finding. Whether the publication was instead of or in addition to informing Cope directly, it doesn't matter. Cope was so upset that he tried to get hold of and destroy every copy of the report.

Geologist Walter H. Wheeler has made a strong case that the break really occurred in 1872. Writing about the feud in a 1960 *Science* article, he argued that Marsh and Cope were both collecting in the Eocene beds of Bridger Basin, Wyoming, in the summer of 1872, and that their competition that summer ultimately resulted in the break.[12]

A couple of letters exchanged by the two men in the following January adds some support to Wheeler's idea. Marsh wrote to Cope, accusing him of withholding some fossils that he considered rightfully his: "The information I received on this subject made me very angry, and it had come at a time I was so mad with you for getting away with Smith [a fossil hunter who had originally worked with Marsh] I should have 'gone for you,' not with pistols or fists, but in print. . . . I was never so angry in my life." Then, curiously, he added, "Now, don't get angry with me for this but pitch into me with equal frankness if I have done anything you don't like."[13]

Was he still hopeful that a rapprochement was possible? In response, he received, not rapprochement, but the frankness he requested: A couple of days later, Cope responded with various complaints such as, "All the specimens you obtained during August 1872 you owe to me."[14]

The scientific community began to hear of the growing enmity in the pages of the *American Journal of Science,* to which Marsh had good access, as well as the *American Naturalist,* which Cope eventually bought in 1877. The accusations in these journals centered mainly on dates of publication and on the accuracy of interpretation. Date of publication usually establishes priority, but publication can take time, and both men tried to use the dates on which they had shipped off the specimens. In the *American Naturalist,* for example, the result was a mess of contradictory dates and attributions, some of which had to do with the speed at which Cope was doing his own work, and some with the fact that he was busy in the field and not on hand to supervise publication. Marsh, however, chose to believe that Cope was deliberately falsifying dates of publication in order to achieve priority.

Still, both men continued to maintain a semblance of civility in public. As late as 1877, Marsh could still state in print, "The energy of Cope has brought to notice many strange forms, and greatly enlarged our literature."[15] Privately, however, Cope, in commenting on the effort Marsh put into his corrections, referred to Marsh as the "Professor of Copeology at Yale."[16]

Out in the Open

In the spring of 1877, Marsh was sent a huge vertebra by a professor in Morrison, Colorado. At almost the same time, Cope received equally impressive bone fragments from a schoolteacher in Cañon City, Colorado. Both Cope and Marsh rushed into print, announcing the discovery of the largest land animal yet found. Each hired the men who had sent the specimens.

These activities were only prelude to the main field event, which was to take place in Como Bluff. Marsh's forces got there first. Williston, one of his men, wrote to Marsh that the bones "extend for seven miles and are by the ton."[17] The year 1877, in fact, marked both the beginning of dinosaur discoveries on a scale unmatched before or after in North America and the mounting of highly successful searches in wild, remote country by relatively well-supplied expeditions of trained people.

The feeling of need for deception at Como Bluff started immediately. Williston wrote to Marsh from his home in Kansas: "It is impossible almost to keep my movements here unknown so that I shall give it out that I am going from here to Oregon! and nobody will know where I am."[18] He also took with him a list of coded words to designate both specific fossils and even the names of Cope's men—who were trying to horn in on the finds—so that when he sent telegrams back to New Haven, he could preserve secrecy. Cope himself eventually did visit, in 1879, and while again a semblance of civility was maintained, Marsh's teams did everything they could to mislead him.

Cope also later accused Marsh of tying up land under the very generous terms of western occupation laws, so as to keep others from entering. Right from the first, however, the claims going back home from both men were creating problems. In 1877, Cope named a new dinosaur *Dystrophaeus viaemalae,* or wasted one. He also claimed this discovery to be the first find of a complete dinosaur in North America. He ignored, knowingly or not, the fact that an earlier dinosaur discovery, found in 1858 by Ferdinand V. Hayden in Haddonfield, New Jersey (Cope's old stomping grounds), had already been described by Leidy. Hayden later supplied reports for Cope, which Cope found useful in preparing his extremely influential *Vertebrata of the Tertiary Formations of the West* (1885).

As noted previously, Marsh had essentially retired from field-work in 1874, much earlier than had Cope, who continued to go out with his expeditions. This left Marsh more time to devote not only to his collections, but also to his own brand of politicking. Elected to the presidency of the National Academy of Sciences, he used his influence to get Cope removed from the U.S. Geological Survey (USGS). This organization was very useful to Marsh's expeditions, as it could have been to Cope's.

Heralding the News

As a result of this and other actions of the USGS, Cope included among his targets in the *Herald* attack John Wesley Powell, director of the USGS. Marsh and Powell worked very closely together for 14 years, 10 of them during Powell's leadership of the organization. Cope charged in blistering terms that Powell reneged on a promise of funds to enable him to carry out publication of his next major work, the continuation of his *Vertebrata,* which the USGS had supported, and which had received plaudits worldwide. (The next volume, with its many complex illustrations, would in truth have cost a small fortune to publish.)

Powell was also insisting—at Marsh's instigation?—that Cope first turn over his fossil collection to the government before the survey would fund the publication. Cope maintained that he had obtained the fossils with his own funds. He also charged

Powell with using his position for personal advantage—such as putting relatives on the payroll.

Cope's gripe against Powell was basically about money. From Marsh, he wanted blood. One of his most serious charges involved plagiarism, and he put forth written charges from disgruntled former employees of Marsh's that much, if not all, of the scientific output credited to Marsh was the work of others under his employ. This was an obvious exaggeration, yet Cope was able to produce several written statements backing up the claim. Williston, who had left Marsh's employ in 1885, had not been reticent about putting his complaints in writing, including a letter to Cope, in which he claimed that Marsh's many published papers were "either the work [of] or the actual language of his assistants."[19]

The letter had gone into Cope's treasure trove of anti-Marshiana and was quickly supplied to the *Herald*. By this time, however, Williston may have had second thoughts about having been so free with his complaints and expressed unhappiness in a subsequent issue of the newspaper, saying that the letters to Cope were private and were never intended for publication. Cope later called him a coward for not standing up for what he obviously felt.

The intrigues and backstage activity must have been a show in itself. Among other people Cope advanced as supporters of his position was George Baur. Although Baur had actually been treated somewhat more generously by Marsh than had many of Marsh's other employees, he felt that he had deserved better, including a professorship. But he got no support from Marsh; his unhappiness was well known.

Baur, however, not only was still in Marsh's employ at the time of the *Herald* series but was also in debt to him, having borrowed some money he had yet to repay. He therefore sent a note to Marsh, who forwarded a copy to the *Herald*, stating, "I have never in any way authorized the use of my name in any attack on you [Marsh] or your work."[20] Cope countered, in print, that Baur had been pressured by Marsh into sending this note, and Baur did resign right after the series appeared.

Marsh, of course, had his own supporters. For instance, George Bird Grinnell, a student of his whom he later employed, remained a staunch supporter. It's worth noting, however, that Grinnell's major contact with Marsh was out in the field, where and when Marsh was at his best.

And so it went. Marsh probably did not help his cause any when, buried in his replies, he referred to some of his accusers as "little men with big heads."[21]

The final installment of the series, dated January 26, 1890, contained a long letter from Otto Meyer, a German who had worked for Marsh from 1884 to 1886. Meyer threw out a series of damning charges about Marsh's methods. In his conclusion to his letter and in a fitting climax to the *Herald* series, he noted, "I presume that all true scientists have more regard for a little man with a big head than for a big man with a little head."[22]

Nor was even this the end. The main storm was over, but rumblings continued. Another player who had refused to enter the ring during the main event was John Bell Hatcher, even though Cope cited him as one of Marsh's unhappy employees. Hatcher, who had made some spectacular finds for Marsh, including the bizarre *Triceratops,* was at that very time out on a dig for him. Although Hatcher had indeed complained about slow payment of funds while out in the field, and also about Marsh's media clamp, he nevertheless stayed on with Marsh and was finally given permission to publish in paleontology under his own name in 1891—after the *Herald* series! He only left Marsh's employ in 1892, after Marsh's USGS funds were finally cut off, and went on to a distinguished career of his own.

Later, referring to one of Marsh's claims in one of Hatcher's own publications (1903), he wrote with dry irony: "In a total of three and one-half days field work he seems to have found sufficient time to 'carefully explore' the geological deposits of the *Ceratops* beds and to trace them for 'eight hundred miles along the eastern flank of the Rocky Mountains,' besides making numerous other observations of scientific interest."[23]

Fallout

As a result of the feud, Marsh felt compelled to take on far more tasks than he really had time for, and as a result, he failed to leave finished manuscript (whether generated by his own efforts or those of others) for any of the several comprehensive monographs he had hoped to produce. A further result is that after his death in 1899, much of the fossil record that he knew so well had to be restudied by others who had to work without benefit of his knowledge.

Cope, no doubt, would not have felt the need to gamble his funds away on fruitless mine adventures, had he not been feuding with Marsh. He died in 1897. Compared with Marsh, his last days were sad ones. During the difficult financial period, he had to sell his house and ended up living in his museum. One biographer, Url Lanham, writes that Cope "spent his [final] illness on a cot surrounded by piles of bones. . . . Six men sat quietly around his coffin at his Quaker funeral, amidst the fossil bones, with a pet live tortoise and Gila monster moving stealthily about the room."[24]

The hostilities had another unfortunate outcome. Joseph Leidy, who had been an early and important worker in this exploding field, had also visited one of Marsh's sites in 1872. The very model of a timid professor, he seemed to pose no threat to Marsh and his forces, so they had no fear of his visit, as they had with Cope's. But Leidy was so put off by all the skulduggery and feuding that, shortly after this visit, he simply left the field of paleontology altogether and turned to other studies.

A subheading in the January 14 issue of the *Herald* read: "Like the Kilkenny Cats, if the Squabble Much Longer Continues There Won't Be Much Left of Either Combatant."[25] Fortunately, the squabble did not kill either combatant. The charges against Marsh were obviously exaggerated; the scientific names proposed by him are still in use for four of the six suborders of dinosaurs recognized today. He was perhaps the first influential supporter of evolution on this side of the Atlantic, and he produced a lineup of horse fossils that showed a powerful (though later discovered to be somewhat flawed) display of evolution in action. Even Hatcher, who did not hesitate to point out Marsh's tendency to exaggerate his own participation, nevertheless also praised him in other publications, both as a person and as a theorist.

There were other happy consequences. The rivalry may have taught a lesson to those who followed Cope and Marsh into the world of paleontology. Certainly, subsequent workers in the field have found it possible to work without feuding. In a much later expedition outfitted by the Carnegie Museum, it was only after the group decided to discontinue operations at a large quarry in Utah that other groups moved in. This wonderful area is now part of what has become known as Dinosaur National Monument, where displays actually show dinosaur fossils in situ. Today, if a group finds and works a promising lode, others will not try to horn in on it. Even better, cooperation suddenly became acceptable. In another expedition supported by the Union Pacific Railroad, a large expedition with members of several major groups all worked together harmoniously.

The happiest consequence, however, is that the pioneering efforts of these two men provided a solid base for what came later. Now let's put some numbers to this claim. To put it simply, the results of the competition were no less than astounding, particularly in light of the primitive search and excavation methods in use. For 10 years, Marsh's forces alone shipped back to New Haven an average of a ton of fossils each week! Cope found other fields, equally productive, and sent back great quantities of specimens as well to his own collection in Philadelphia.

The finds, the news of them, and their display as complete creatures in museums and exhibitions sparked the beginning of the public's love of these marvelous, and wonderfully varied, creatures. As Colbert has put it, "the dinosaurs came alive during those last two decades of the nineteenth century."[26] Compare Colbert's statement with one that appeared in *Blackwood's Edinburgh Magazine* in 1855, in which the anonymous author pointed out how "we dread the name of Museum, and tremble at the sight of a collection of specimens."[27] The new discoveries also encouraged a variety of purse holders to help support the increasingly expensive expeditions needed to get at the ever-less-available specimens being sought.

It wasn't only the dinosaurs that created excitement in the world of paleontology, however. The very first quarry at Como Bluff gave up a tiny mammal jaw that turned out to be the first Jurassic mammal found in North America. It was suddenly realized that even here in dinosaur country, small fossils could be as important as large ones. After all, the history of mammals is surely as important as the history of dinosaurs.

It's probably not too great a stretch to argue that even today's broad-based support for science dates back to the Cope–Marsh feud. The late Carl Sagan, one of the world's great popularizers of science, once stated in an interview: "Much of the funding for science comes from the public. . . . If we scientists increase the public excitement about science, there is a good chance of having more public supporters."[28]

Talk about excitement: Between 1877 and the late 1890s, Cope, Marsh, and the teams they sent out had excavated, studied, characterized, and named nearly 130 new species of dinosaurs, including the menacing *Tyrannosaurus,* the monstrous *Brachiosaurus,* and the grotesque *Triceratops.*

Now, a century later, the excitement—rather than dying down—is heating up. Not only are present-day fossil hunters still making new fossil discoveries, but also the vast trove of fossil treasure that has already been collected is being looked at in the light of new perspectives and updated theories about the evolution of animal life, especially that of dinosaurs.

The word *dinosaur,* for example, has for a long time carried the meaning of out-of-date, impracticably large, a leftover. That may have to change, for as a group, dinosaurs are now more likely to be viewed as capable and intelligent and perhaps quite as fast-moving as the broad-based mammalian group is today.

If they were so capable, however, why did they die out? Determining the causes of their extinction has become an industry in itself and has drawn a whole phalanx of different disciplines into the search.

Better methods have also been developed for finding specimens. In one demonstration, Los Alamos scientists were brought in and tried sound waves, radar, highly sensitive chemical tests, and even a night search using ultraviolet light.[29] Contemporary paleontologists have more efficient digging tools and use helicopters and other advanced transportation methods to carry out specimens from wild and inconvenient places.

Nevertheless, many of the most awe-inspiring dinosaur displays the world over can be traced back in some way to the intense feud between Cope and Marsh, which drove them to sometimes superhuman effort. Some of their theoretical contributions have been important as well. One, Cope's Rule, has been a standard organizing principle in paleontology since he propounded it more than one and a quarter centuries ago. The rule states that all species, from fungi to whales, tend to increase in size over time. A recent comprehensive analysis suggests, however, that while it is certainly true for some species, it does not hold up in all cases. Marsh would have been delighted.

CHAPTER 8

Wegener versus Everybody

Continental Drift

Early in this century, Alfred Wegener, a young German scientist, proposed his theory of continental drift. The basic idea was this: All of Earth's continents were, at some time in the distant past, joined into a single great land mass, which he called "Pangaea," and the various parts we see today broke apart some 200 million years ago and have been sailing majestically across Earth's surface, like huge icebergs over a denser substratum.

Today, we have little problem with his idea; in fact, it is the foundation on which all of modern earth science is based. When Wegener put it forth, however, the reaction was not only negative, but so intense that many who might have found themselves in his corner hung back in fear of endangering their own careers. For five decades, its few proponents were dismissed contemptuously by scientists on both sides of the Atlantic, but most strongly by those in the United States. Critiques often included such words as *preposterous, antiquated, serious error, footloose,* and even *dangerous.*

The reasons for the dismissal of both the idea and its author are many and instructive. It may have been partly the idea's connection, however tenuous, with catastrophism, which was out of favor at the time. Today, we know that Earth's history incorporates elements of both catastrophism and uniformitarianism. Thus, Lord Kelvin was intuitively correct in supporting catastrophism, *and* Thomas Henry Huxley, on the side of the geologists, had reason to support uniformitarianism.

Some of the reaction also had a kind of not-in-my-back-yard flavor, for Wegener—an astronomer and meteorologist—was an outsider as far as the earth scientists were concerned. And in truth, continental drift was of peripheral importance to him. His own father-in-law, a respected meteorologist in his own right, was one of Wegener's earliest critics and tried to dissuade him from straying out of his own field of expertise.

So strong was the reaction that the name of Galileo was invoked more than once by his few supporters. In 1926, for example, Harvard's Reginald A. Daly published a book called *Our Mobile Earth;* on its title page appeared the words *E pur su Muove* (roughly, "But it does move"), which Galileo is said to have mumbled under his breath after his humiliating abjuration.

Although the literary allusion to Galileo is a strong one, Wegener's case has far more in common with that of Darwin. In fact, a fascinating set of parallels between the two situations practically tells the story.

A Strong Parallelism

Wegener (1880–1930), like Darwin, was born and brought up in comfortable circumstances. Driven by dreams of exploring northern Greenland, he built up his endurance with long days of walking, skating, mountain climbing, and skiing. Vigorous, healthy, and daring, he participated in some audacious adventures, including (with his brother Kurt) a balloon flight of more than 52 hours. This flight set a record and was an especially bold venture, considering the primitive equipment available at the time.

Both Wegener and Darwin undertook long, difficult expeditions in their young lives and did extensive data collecting— Darwin mainly during a five-year voyage on the good ship *Beagle,* and Wegener during several long sojourns in Greenland. In 1913, Wegener's expedition was threatened during the ascent of the inland glacier by major calving (splitting) of the ice, extending right up to the camp. The group's crossing of the island lasted two months and was completed only with the greatest difficulty.

Like Darwin, Wegener was trained in an area of study that had little to do with the topic in which he made his mark. Darwin had studied medicine and theology and did early scientific work in geology. Wegener took a doctorate in astronomy and became a practicing meteorologist. After returning from his first sojourn in Greenland (1906–1908), he became a lecturer in both astronomy and meteorology at the University of Marburg, in Germany. He is reported to have been a fine, popular teacher.

Brave and robust in those younger years, he was apparently also a peace-loving person. This made his service during World War I all the more difficult for him. Also, as with Darwin, he did his major work in the midst of a health problem. Shot twice, he was no longer fit for active duty and was transferred to full-time employment in the military's field meteorological service. Further, although he first presented his ideas on continental drift in an article and lecture in 1912 (before World War I), his fame rests on the book he wrote during the war (published in German in 1915). This means that he somehow managed to come up with a major, earth-shattering book while on sick leave and while finishing out the war in the field meteorological service.

His book was titled, significantly, *The Origin of Continents and Oceans.*[1] Like Darwin, he used the term *origin* in the title, and he dealt, essentially, with evolutionary concepts.

And in both cases, the basic concept also cut across many lines. "The book," Wegener wrote in his foreword, "is addressed equally to geodesists, geophysicists, geologists, palaeontologists, zoogeographers, phytogeographers [phyto = plant] and palaeoclimatologists. Its purpose is not only to provide research workers in these fields with an outline of the significance and usefulness of the drift theory as it applies to their own areas, but also mainly to orient them with regard to the applications and corroborations which the theory has found in areas other than their own."[2]

In other words, like Darwin, he had brought together evidence from a wide diversity of fields. Therefore, he and his few followers were sparring with a whole range of opponents, each of whom saw him as an interloper. At this time, for example, the idea of a cooling, contracting Earth was still held by most geologists, who felt that such cooling and contracting was the only possible explanation for a variety of observations, including mountain building. Like a rotting, shriveling tomato, they thought, a contracting Earth produces both peaks and valleys in its skin. Wegener pointed to the discovery of radium and argued that the idea of a cooling Earth no longer made sense, offering in its place his own idea that the movement of continental masses slowly mashing together at some point in the past was a better explanation of mountain building.

Nevertheless, he, like Darwin, recognized the weaknesses in his theory, so he too put out several editions of his book, each heavily revised on the basis of new information and criticism. In Wegener's fourth revised edition (1929), he was still writing, "In spite of all my efforts, many gaps, even important ones, will be found in this book."[3]

Like Darwin, he was not the first to come up with his theory. In Wegener's case, the broad outline of the idea had been tentatively broached several times. Wegener wrote, "I encountered many points of agreement between my own views and those of earlier authors." One of those he mentions is H. Wettstein, who in 1880 "wrote a book in which (besides many inanities) the idea of large horizontal relative displacements of the continents is to be found. . . . However, Wettstein . . . regarded the oceans as sunken continents, and he expressed fantastic views, which we pass over here."[4]

The jigsaw fit of the continents was obvious and had been noticed as far back as the sixteenth century, when the first reasonably accurate maps began to be drawn of the New World. Francis Bacon is usually given priority credit for seeing the rough congruence, which he mentioned in his great *Novum Organum* of 1620. Actually, he was only commenting on the similar shapes of South America and Africa. In 1994, a professor of classics at Bard College, James Romm, traced the drift lineage back to a Dutch cartographer named Abraham Ortelius; according to Romm, Ortelius proposed the idea in 1596.[5]

It was Wegener who put meat on the idea, however, and developed it into something that could not be ignored. As with

Darwin, then, Wegener's *Origin* became an in-the-face kind of thing. The result was a theory that was not only dramatic but also so all-embracing that few of his opponents felt they could attack it as a whole. Therefore, they went after it piecemeal, depending, of course, on their own interests and disciplines.

Details, Details

The details were indeed attackable, for at the time, far less was known about Earth than is known today. The ocean depths, covering fully 70 percent of Earth's surface, were a deep, dark secret. Sonar, deep drilling, and many other modern techniques lay in the future, so the regions far below the continental lands were just as mystifying as those beneath the oceans. As a result, Wegener had to guess about many of these details. But he felt that the grand sweep of his idea was what counted.

Today we are impressed with the audacity of a single hypothesis that could explain not only the origins of mountains and oceans, but also many of the puzzles that confronted the various specialists Wegener mentioned in his foreword. Among these puzzles were intriguing similarities on both sides of wide oceans—similarities in rock formations, as well as in living things past and present. Another was a distribution of climates in the past that was different from today's, shown for example by glacial leftovers in Africa and evidence of tropical species in the far north and south.

In Wegener's day, however, a zoogeographer looking at similarities in living things in Africa and South America might never hear of equally strange correspondences in rock formations that would intrigue geologists. Wegener, the outsider, could take the broader view, and he could see the forest, while the specialists could see only the trees. Yet, as with evolution, after the first flurry of activity, there was a long, slow fading of the theory from view, then a subsequent resurgence of interest in the mid–twentieth century, as new evidence showed the strength and beauty of Wegener's hypothesis.

Mechanism

Another striking similarity between the two cases is that Wegener, like Darwin, was not able to propose a satisfactory mechanism for his theory. With Darwin, you recall, it was the genetic aspects of selection that were missing and were not found until later.

Although Wegener knew he was right, and that everything pointed in the right direction, he too could not come up with an appropriate mechanism—in his case for the grand march of the continents. What kind of engine could drive continent-size masses through a rocky underlayer, soft or not?

The best he could come up with was a pair of candidates. One was what he called *polflucht,* pole-fleeing forces; these were due to the rotation of Earth and caused a migration of the continents toward Earth's equator. The second, which had to do with lateral displacement, he attributed to some sort of tidal drag due to gravitational forces from the Sun and Moon.

He suspected that these forces were not powerful enough to drive continents and fold mountains, but he felt that they could perhaps suffice if acting over long periods of time. And he had nothing better. Modest and sensible, he admitted that "The Newton of drift theory has not yet appeared."[6]

This weakness was to provide strong ammunition for his critics. Harold Jeffreys, whose influential book *The Earth, Its Origin, History and Physical Constitution* (1924) established a firm mathematical basis for geophysics, did some calculations and showed that the pole-fleeing and tidal forces were about one-millionth as strong as would be needed to displace continents and produce mountains. He also produced a complex quantitative theory of cooling and differential contraction that, he claimed, provided the needed forces. The logic, à la Kelvin, was unassailable and effectively stifled drift theory for decades.

Finally, like Darwin, Wegener could not prove his theory. Broad-ranging theories such as evolution and continental drift are, by their very nature, difficult to prove. A geological theory is especially unlikely to submit to conventional laboratory experimentation, and even field observations are difficult because of the great time and space requirements. As a result, Wegener could present only indirect evidence.

Wegener thought he had some direct evidence when he compared historical readings of Greenland's longitude with new ones taken in his day. Unfortunately, the measurements were not up to the task, as was easily shown by his critics.

Differences

No analogy is perfect, however, and there were some surprising differences between Darwin's and Wegener's situations. A major one had to do with time and preparation. The first known notice that Wegener took of the jigsaw-puzzle fit occurred in 1903, when he mentioned it to a fellow astronomy student. What happened next is best explained in his own words, written some years later:

> The first concept of continental drift . . . came to me as far back as 1910, when considering the map of the world, under the direct impression produced by the congruence of the coastlines on either side of the Atlantic. At first I did not pay attention to the idea because I regarded it as improbable. In the fall of 1911, I came quite accidentally upon a synoptic report in which I learned for the first time of palaeontological evidence for a former land bridge between Brazil and Africa. As a result I undertook a cursory examination of relevant research in the fields of geology and palaeontology, and this provided immediately such weighty corroboration that a conviction of the fundamental soundness of the idea took root in my mind.[7]

He first presented his idea in two addresses in January 1912, just 4 months after seeing those reports. Even if, to keep the field even, we wait until publication of Wegener's *Origin,* the time span between eureka and publication is only 5 years, a far cry from the 20 years Darwin took to produce his masterpiece. In addition, although Wegener's relative unfamiliarity with the various fields involved may have enabled him to see their weaknesses more easily, it also meant that he was to present the

idea without an inkling of the storm that was to fall on him—
again, a far cry from Darwin's fears and hesitations.

Another difference is that the attacks on Wegener were not
religiously based. This lack of religious fervor could explain why
the theory of continental drift is today recognized as a powerful, if incomplete, picture of Earth's evolution, while Darwin's
evolution continues to be attacked as untrue, at least by fundamentalist groups.

Pet (and Pat) Theories

When Wegener first presented his thesis in 1912, the scientific
world was only just extricating itself from the bonds of Kelvin's
confining age-of-the-Earth calculations. This relative freedom
provided a great deal more room for speculation on prehistoric
conditions on our planet, and interest was more intense than
ever. But the idea of a cooling, contracting Earth remained a
potent one.

The early twentieth century was also an era when broad
areas of earth science were thought to have been established on
solid scientific grounds. Few welcomed an idea that overturned
almost everything then believed about the earth sciences. As
late as 1928, the American geologist R. T. Chamberlin could still
write, "If we are to believe in Wegener's hypothesis we must for-
get everything which has been learned in the last 70 years and
start all over again."[8]

One of the things that had been learned was that something
had to bridge the wide Atlantic Ocean in order to explain the
increasing number of reports noting similarities being found on
both sides. An important example of these reports concerned
Glossopteris, a well-preserved fern unearthed in coal beds of
the late Paleozoic era (roughly 250 million years ago). Beautifully preserved fossils of this fern had been found over the years
in areas as widely separated as India, South Africa, Australia,
and South America. It just didn't make sense to believe that the
same species of this fern could have arisen independently in
these various areas. Some connection was needed.

Well, if two continents such as North and South America can be connected by a land bridge, why couldn't the same be true for, say, South America and Africa? The only difference would be that the bridge between North and South America has remained in place, while the other has subsided over the course of time. Another idea that achieved some prominence was that continent-size masses had once existed but had been buried or had sunk. The land bridge seemed the most likely choice.

In other scientific areas, other ideas came into being. By the end of the nineteenth century, extensive gravity surveys added strength to an idea called "isostasy," which suggested that mountains and the crust beneath them comprised material that was less dense than that of the ocean floors. If both the continents and the ocean basins were floating on a denser substratum, it is easy to see that the lighter continental portions would float higher in the underlying material. If the bases of mountains have the lowest density of all, this lightness would explain why mountains rise above the rest of the land areas. It was known that vertical movements do take place: Careful observations had shown that Scandinavia had sunk under the weight of glaciers during Pleistocene times and has been rising again in the warmer postglacial climate. If isostasy explained mountain-building, it thereby supported another powerful idea of the time: the permanence of the world's layout of continents and oceans. With permanence comes an argument against a contracting Earth.

Facing these factors, Wegener stated in his book: "So here we have the strange spectacle of two quite contradictory theories of the prehistoric configuration of the earth being held simultaneously—in Europe an almost universal adherence to the idea of former land bridges, in America to the theory of the permanence of ocean basins and continental blocks."[9]

"Where does the truth lie?" he continued. "The earth at any one time can only have had one configuration. Were there land bridges then, or were the continents separated by broad stretches of ocean, as today? . . . There clearly remains but one possibility: there must be a hidden error in the assumptions alleged to be obvious."[10]

If vertical movements of major portions of Earth are possible, then why not horizontal ones too? Wegener threw out his challenge, one that brought down the wrath of scientists, not only from varying disciplines, but also from different sides of the Atlantic.

Objection!

The first German edition of Wegener's *Origin,* only 94 pages long and lacking an index, did not excite much interest. Four years later, in 1919, another German edition appeared; this one—better organized, with more evidence, and with an index— caught the attention of scientists on the European continent. Colleagues in the United States remained blissfully unaware of the storm that was brewing until the third edition (1922) was translated into several languages, including English.

Two important geologists—the Briton Philip Lake and the American Harry Fielding Reid—then wrote critical reviews, and at this point the chorus of attacks rose to shrill levels, including some that questioned Wegener's credibility as a scientist. Lake complained, "he is not seeking truth; he is advocating a cause, and is blind to every fact and argument that tells against it."[11] Lake also argued, "It is easy to fit the pieces of the puzzle together if you distort their shape, but when you have done so, your success is no proof that you have placed them in their original positions. It is not even a proof that the pieces belong to the same puzzle, or that all of the pieces are present."[12]

The Americans also went after Wegener in earnest. Paleontologist E. W. Berry called his theory "a selective search through the literature for corroborative evidence, ignoring most of the facts that are opposed to the idea, and ending in a state of auto-intoxication in which the subjective idea comes to be considered as an objective fact."[13] American geologist R. Thomas Chamberlin (son of one of Kelvin's powerful adversaries) wondered whether geology can still be considered a science if it is "possible for such a theory as this to run wild."[14] Bailey Willis, another respected American geologist, maintained that "further discussion of it merely incumbers the literature and befogs the mind of fellow students. [It is] as antiquated as pre-Curie physics."[15] He also called it a "fairy tale."[16]

Among the strongest objections were those raised by geophysicists. Wegener had argued that the continents consisted of a rocky material called "sial," and that these were gliding along on a denser but softer substratum, which he called "sima." He explained the softness by assuming that sima melted (i.e., attained a fluid state) at a lower temperature than sial. Unfortunately, physical experiments contradicted his conjectures regarding the melting point of sima. In addition, observations of seismic waves showed the ocean floor to be rigid, not soft, so his theory was taken to be unscientific.[17]

We know now that Wegener was nevertheless right in concept. Oddly, as this chapter subsequently shows, he was thinking not too big, but too small. Recall his hypothesis that pole-fleeing forces and gravitational forces, acting over long periods of time, could perhaps cause continental drift. Wegener was a good enough scientist to recognize these forces as a major problem in his theory, so that the mechanism for continental drift was a weak link in his theory.

Among those who challenged this aspect of his work was Harold Jeffreys, mentioned earlier in this chapter. Jeffreys called the drift idea "a very dangerous one, and liable to lead to serious error."[18]

Helpless against the onslaught, Wegener could only complain. To his father-in-law, he wrote, "Professor P.'s letter is typical! He will not allow himself to be taught. Those people who insist in treating only with the facts and want to have nothing to do with hypothesis, themselves are utilising a false hypothesis without appreciating it! . . . there is nothing in his letter about the struggle to get to the bottom of things, but only about the pleasure of exposing the limitations of other men."[19]

Wegener believed that the individual potshots taken at the drift hypothesis could not bring it down. He was wrong there. On the other hand, he also believed that the truth of the theory could be established only by combining all the evidence. He was one of the very few who seemed able to do so.

Allies

Wegener, although utterly outgunned, was not completely alone. In 1928, about two years before Wegener died, an Edinburgh University geology professor, Arthur Holmes—who did important work with geological dating—calculated that volcanic activity was not sufficient to account for the heat generated by radioactivity. He thereby postulated the idea of thermally driven convection currents in the earth. The process was similar to that in a large pot of boiling water. Heat generated by the burner at the center of the pot's bottom will produce currents of water moving up from the bottom and out to the circumference. Here then was a potentially appropriate motor for continental drift, and Wegener immediately included it in his 1929 edition of *Origin*. Unfortunately for Wegener, although such convection currents could serve as a motor, it was still not clear just how this would happen. Also, there remained plenty of other unsolved problems, so the idea, while basically correct, was not very helpful.

At about the same time, Alex du Toit, a South African and one of the most important field geologists of that time, had noted a striking similarity between the Paleozoic and the Mesozoic geology of his native country and that of eastern South America. He gathered other evidence and became an enthusiastic advocate of continental drift. Wegener included some of this evidence in the 1929 edition of his book.

But there still was nowhere near enough evidence to lift Wegener's theory from the depths of controversy. The important question of mechanism still had no satisfactory answer. Even many of those who had been converted to drift theory felt it was better to keep quiet and await further developments. In fact, attitudes seemed in some ways to harden against the theory.

As late as 1943 (13 years after Wegener's death), the American paleontologist George Gaylord Simpson remarked on the near unanimity of feeling against it among his peers. He himself argued, "The known past and present distribution of land mammals cannot be explained by the hypothesis of drifting continents . . . the distribution of mammals definitely supports the hypothesis that the continents were essentially stable throughout the whole time involved in mammalian history."[20] As late as 1950, T. W. Gevers, a disciple of du Toit, would refer to a "marked regression away from continental drift."[21]

Turnabout

During the laying of the transatlantic cable in the mid-1800s, a curious formation was discovered in the seabed, about halfway between the coasts of the New World and the Old World. Known as the Mid-Atlantic Ridge, it turned out to be part of a long, submerged mountain range that roughly paralleled the coastal forms of the two coastlines it rested between.

Wegener knew about this submerged range but did not feel it had anything to do with his hypothesis. In his description of the continents' movements, he said it made no difference where the center of movement was placed, that it was the *relative* movement that was important. He laid out three possibilities— with Africa, the ridge, and South America all being possible centers from which the other landforms were moving away.

In terms of relative movement, he was right. If a rubber band with two markings on it is stretched, and you want to measure the new distance between the markings, it makes no difference which mark is chosen as the zero point. In terms of drift theory, however, it mattered a lot which landform was chosen as the center. The answer did not come easily.

With the advent of World War II, a series of major improvements in equipment and techniques for mapping led to many discoveries. These discoveries not only provided an answer to the landform question, but also, finally, changed the theory's fortunes. These revelations came about in two quite separate areas.

The first discovery, derived from deep-ocean mapping, showed that the Mid-Atlantic Ridge was just one of many such ridges. The series can be described as a worldwide, submerged mountain chain, but one that is quite different in shape and constitution from any found on land.

Ridges are, in fact, found in all the oceans. They also form a sort of continuous seam, like that on a baseball. Further, the seam is dotted here and there with submarine volcanoes and scattered volcanic islands; such islands include the Galapagos Archipelago, Ascension Island, and Iceland. The hottest and youngest areas of the ridges, significantly, are found near the center lines of the ridges.

Scattered observations in the world's seas, like clues in a murder mystery, continued to pile up. Here's another clue: As dating techniques were refined, they showed that no part of the seafloor is more than 200 million years old, which is far less than the age of the continental rocks. This discovery was a real shock; the traditional view, remember, was that the ocean floors and the continents had all been created at the same time. Studies also showed that (a) the continental crust is made of different material from ocean crust, (b) the ocean crust is much thinner than the crust underlying the continents, and (c) a denser material underlies *both* ocean crust *and* continental crust.

The second area of research that changed the fortunes of drift theory had to do with magnetic information sealed into Earth's rocks over its long history. By the late 1950s, data from years of magnetic readings taken with magnetometers towed behind ships showed several surprising phenomena. One was a strange pattern of magnetic stripes along the ocean floors. They appeared symmetrically along both sides of, and more or less parallel to, the ridges, and with alternating polarity. These stripes were particularly puzzling, but understanding was not long in coming.

Seafloor Spreading

In 1960, Harry H. Hess of Princeton University advanced an idea that, like Wegener's, integrated information from a variety of sources. It was simple and brilliant: The seafloor comes into being at the midocean ridges, rising from Earth's depths as hot, ductile lava (or magma). Like a new, long volcano rising from inside Earth, the material builds up as it emerges, forming the great mountain chain that rises miles above the ocean bottom. The magma also spreads out in two opposite directions, away from the ridges, forming new ocean floor. In all known cases, these ocean floors are no more than 200 million years old.

The idea, at first, had little more effect than Wegener's did at the beginning—but help was coming. Thanks to work by several other scientists,[22] the alternating magnetic stripes could be seen as a kind of fossilized magnetic tape. As the molten-rock material emerges and cools, the orientation of Earth's magnetic field is locked into this material.

The worldwide magnetic field is known to have reversed itself many times during Earth's long history, and the alternating stripes show the direction of the field at the time the rock emerged and cooled. As the material was pushed away from the center, it retained this orientation, and, as new material emerged after a reversal, this new material showed the opposite polarity. Clearly, large portions of Earth's surface were moving. Here, then, was good supporting evidence for the continental drift hypothesis.

Hess's idea came to be called seafloor spreading. Among yet other puzzles that it solved was why the lava at the crest always seems to be younger than the lava lying farther away from the center. Most important for our story, however, Hess's idea provided a satisfactory, and powerful enough, motor for Wegener's continental drift. The continents are carried along for the ride by the overall process, which is driven by convection currents in the *mantle* (the thick layer between Earth's crust and its core deep within). Hess spelled out the difference thus: "The continents do not plough through oceanic crust impelled by unknown forces, rather they ride passively on mantle material as it comes to the surface at the crest of the ridge and then moves laterally away from it."[23]

At this point, continental drift was back in business. Not a complete answer in itself, it became part of another growing theory, called "plate tectonics." What the modern synthesis has done for evolution, plate tectonics has done for earth science.

Plate Tectonics

In this new scenario, the continents are not ships at sea, sliding through Earth's crust. Rather the outermost portion of Earth is divided up into a series of hard, rigid plates that vary in thickness. According to the most recent belief, the plates include not only Earth's crust, but also a portion of the upper mantle. Under the oceans, the plates range in thickness from a mere 4 or so miles to perhaps 80 miles in the oldest parts of the ocean floor; the continental plates are generally much thicker, ranging from about 20 miles down to a depth of some 180 miles below Earth's surface.

This grouping of plates, the outer layer of Earth, is known as the "lithosphere" (*lithos,* from the Greek word for "stone"). The plates are floating on a plastic layer of the lower mantle called the "asthenosphere" (from the Greek *asthenēs,* "weak"). These gargantuan plates, which may or may not correlate with the continental margins, are moving about on Earth's surface, driven by slow but powerful currents of molten rock.

Where the edge of one plate meets that of another, all kinds of interesting things can happen. One of the plates may descend back into the mantle; it may ride up on top of an opposing and lighter block; or perhaps it will mash up into a mountain range. The western margin of the United States and the eastern edge of Asia are believed to be the boundaries of moving plates. As these blocks move, they are prone to cracking or breaking up along the edges, which could account for the high concentration of earthquakes and young mountains in these regions. Further, friction created where two blocks meet generates tremendous heat, which can melt underlying strata of rock. The great pressures within Earth force the resulting magma up and out, creating volcanoes and the lava that flows out of them.

Today

Happily for geologists, there remain plenty of unresolved questions in this expanding field. As a matter of fact, the mechanism problem faced by Wegener has not been fully solved even today. Plate tectonics does a satisfactory job in explaining the movement of oceanic crust. It does less well, however, in explicating the movement of continents, which are thicker than oceanic plates and extend deeper into Earth's mantle. A 1995 proposal suggests that the tug of old ocean floor sinking back into Earth is what powers most plate motions.[24]

The search goes on in many areas. One has to do with the bucking contest going on right now between the continental plates underlying India and the rest of Asia. For 50 million years, India has been driving north into the rest of the Asian continent at about 5 centimeters per year. So far, says K. Douglas Nelson, a geologist at Syracuse University, it has "pushed up the Himalayas and the Tibetan Plateau, and pieces of central Asia are pushing out like melon seeds into the Pacific."[25]

In other words, the Indian plate is sliding down under Asia, with the aforementioned results. Recent research adds a twist to the situation, for a kind of melting pot seems to be underlying the region. This was unexpected but may answer a long-standing question: Why is the Tibetan plateau, an area ringed by mountains, so flat? Researchers suggest that the soft underbelly somehow permits the land here to flatten out, just as a viscous fluid such as peanut butter will, if given enough time. Lessons learned from this research can also be applied to a better understanding of earlier collisions.[26]

How about the plates themselves? What determines their size? Theory suggests that they should be no more than about 3,000 kilometers (1,860 miles) wide. Why is the plate under the Pacific Ocean four times that size? New work points to a greater viscosity in the deep mantle than was thought to be the case, which in turn may have a bearing on plate size.[27]

Even the number of plates may be in dispute. The latest count is an even dozen major plates, along with several smaller ones. Recent observations, however, suggest the possibility that the tectonic plate on which India and Australia have been riding together is now splitting apart, which would increase the number of major plates to 13.[28]

With so much still in question, is it any wonder that Wegener didn't get everything right? This complexity makes his basic guess—that Pangaea began to split up about 200 million years ago—all the more amazing, for that's one of the few areas on which everyone seems to agree.

Wegener, throughout his ordeal, managed to continue his own career. In 1919, he received an appointment in the Meteorological Research Department of the *Deutsche Seewarte* in Hamburg, where he was able to combine both civil and academic functions. Five years later, in 1924, he was appointed to a newly created Chair of Meteorology and Geophysics at the University of Graz, in Austria.

Still physically active at age 50, he made plans for a major expedition to Greenland, his fourth, to run from 1930 to 1931. But it ended in disaster; he lost his life attempting to cross from a camp on the central ice cap to the base camp on the west coast. When he died, in 1930, his theory was still in a sort of scientific limbo. His legacy lives on, however—bigger, grander, more comprehensive, and more majestic than even he could have imagined.

CHAPTER 9

Johanson versus the Leakeys

The Missing Link

It isn't often that a scientific feud hits the front page of the austere *New York Times*. On the morning of February 18, 1979, however, there it was, including a three-column picture spread across the bottom of the page. Just under the picture was the headline,

RIVAL ANTHROPOLOGISTS DIVIDE ON 'PRE-HUMAN' FIND

No great drama there. We wonder why this is on page 1. The body of the article begins,

> Two well-known anthropologists challenged each other today in what could become a wide-ranging debate over whether a finding last month was indeed a new species of pre-human being ancestral to all other known forms of human and human-like creatures.
>
> Richard Leakey, the Kenya anthropologist, is challenging the announcement last month by two American scientists that they had discovered such a new species. Dr. Donald C. Johanson, one of the Americans, appeared with Mr. Leakey at a symposium here on human evolution and vigorously defended his interpretation.

"Vigorously?" Were there insults? Fists? Knives? No, nothing like that. What then could have prompted the editors to give this report such prominence? One angle: The up-and-coming down-home American versus the top star in the staid British establishment. It was David and Goliath, with Dr. Johanson, the American, as David. Or was he Goliath? Note too that Leakey only has a "Mr." in front of his name.

Whichever, the report contained none of the explosive charges leveled by Cope against Marsh almost a century earlier. Also, in typical aloof British fashion, Leakey didn't actually jump in with both feet at the Pittsburgh conference. So the story was in the details. And the reporter, Boyce Rensberger, dug up plenty of those. But there was another important reason for the page 1 placement.

By the early part of this century, evolutionary theory, including the evolution of humans, was elbowing its way into the scientific world. The idea, or fear, that we humans are descended "from the apes" was still being bandied about. An alternative, more sensible, theory was that we have evolved from some other, unknown creature that was antecedent to both humans and apes.

There was a real problem with this idea, however—namely, a big hole in the fossil evidence for the human line of descent. We had ourselves at one end of the scale, and we had our cousins the modern apes at the same end. We also had some fossil evidence for ancient apes, thought to date way back to an estimated 10 million years ago and more.

But how about the intermediary stages? Where was the "missing link"? Next to the Holy Grail, the missing link may be the most sought-after prize in human history. Every civilization, every recorded society has myths and legends attempting to explain where we came from. It was about this link, basically, that Leakey and Johanson were wrangling.

The Missing Link

The great Charles Darwin, who has appeared in these pages several times, enters the stage yet again: In 1871, he had predicted that the origins of humanity would be found in Africa. To the ears of early twentieth–century Western Europeans, raised on the comforting idea of white supremacy, those words alone were enough to keep them turned against both Darwin and evolution.

By the time of the *Times* article in 1979, however, a growing list of tantalizing finds and new interpretations had already built up high interest in the missing link among the public. The famous Piltdown Man, for example, created a media sensation. "Found" in 1912, it had a large brain and a small jaw, so it filled with delight a public that saw an enlarging brain as the change that made our ancestors human.

A dozen years later, the Australian-born anthropologist Raymond Arthur Dart came up with another fossil skull in Taung, near the Kalahari Desert, in his adopted land, South Africa. The finder of any new fossil had long had the privilege of choosing how to locate it in the paleontological literature, and of giving it a name. Dart created a new category and named it *Australopithecus* (southern ape) *africanus,* though it was commonly known as the Taung skull.

When he published his results a year later—the same year as the Scopes monkey trial!—the response was explosive, but hardly what he expected. Problem one: He had noted that the *foramen magnum* (a hole in the skull through which the nerve bundle from the rest of the body passes into the brain) was at the bottom of the skull. In contrast, the hole in quadrupeds is located at the rear of the skull. He therefore concluded that this creature had walked upright.

Problem two: Dart's australopithecine skull had the jaw of a human and the brain of an ape, just the opposite of the Piltdown skull. Everybody knew that our large brain was what made us human, so Piltdown clearly seemed the more authentic find.

Problem three: The australopithecine skull was that of a young child, and some critics pointed out that the humanlike aspects could be misleading, that with further development, the basic apelike features would emerge. Dart, they insisted, had found the young skull of some anomalous ape and had made an egregious error.

Still further, the feeling was building that human origins would be found in Asia, the locale for many ancient civilizations.

A contemporaneous find of a fossil tooth in Peking (now Beijing) seemed to support that idea, and Dart's find at Taung just didn't fit. His Taung child created a sensation all right, but mainly as the butt of jokes in cartoons and in music-hall performances.

Dart, disheartened, left the field. Thirty years later, the Taung child was finally recognized as a major find. Part of the reason for its subsequent acceptance was that the Piltdown skull was shown, in 1953, to be a thoroughgoing fake, artfully constructed by persons unknown for some nefarious purpose.

Louis the Tenacious

Dart was defeated by the prevailing views. Another young researcher also had an outrageous idea but was not so easily defeated. In fact, if there is a beginning to the *Times* story, it lies in the life of Richard Leakey's father, Louis S. B. Leakey. Louis, who had spent most of his young life in Kenya, knew by the age of 13 that he wanted to be an archeologist. By 1924, at age 21, he was already participating in a dinosaur-hunting expedition in what is now Tanzania, and he earned money for his schooling by lecturing about the expedition.

This expedition was essentially an introduction to his later career, for this brash young man not only argued that Darwin was right about the African origin of humans, but that he, Louis, was going to prove it. Tall, handsome, and confident, he delighted in goading the academic world as he became better known.

Though he had earned a degree at Cambridge, he was impatient with armchair academicians. He further endeared himself to the academic world by insisting that our human ancestors emerged, not some half a million years ago, as was commonly thought, but much further in the past. Though recognized even in his college years as egotistical and stubborn ("pig-headed" was another designation), he did make friends easily, including Gregory Bateson, ecologist/anthropologist and later the husband of Margaret Mead (discussed in Chapter 10).

In 1926, at age 23, the audacious Louis went on his first search for human fossils. The area he chose to delve in is called the Great Rift Valley, which runs north and south through Ethiopia, Kenya, and Tanzania. At the time, no one thought the region was of any interest; today, it is known to be home to four of the world's richest hominid sites.

The area owes its unusual character to the activities of plate tectonics. Three separate plates come together here, and the shifting plates have buckled the surface, thrust up volcanoes, and formed depressions that became lakes and rivers. This constant, though slow, activity had the effect of continually creating and exposing volcanic and sedimentary layers. Throughout long eras, species rose and fell, leaving fossil evidence of their comings and goings.

Louis zeroed in on his favorite spot of all, Olduvai Gorge, at the northern end of Tanzania. Some 30-plus miles long, with both main and side gorges, it shows clear geological horizons; in some cases, its fossil deposits rise 300 feet above the sandy bottom. But it's a forbidding place, blazing hot, dry as dust. In later years, Richard, who was dragged along on many of the expeditions, recalled, "I remember quite clearly why I never wanted to be a paleoanthropologist. . . . You're always hot, sticky, wishing for shade and swatting at flies."[1] Johanson concurs about the hot part and adds, "I almost always came back with some kind of illness. I had very severe fevers in the 1970s that were never diagnosed."[2]

In addition to the long hours under a scorching sun, other challenges include the logistical problems of supplying food and water for the search party over extended periods of time, hundreds of miles from nowhere; the difficulties of distinguishing stones from bones; and the challenge of having to search large areas for, at times, the tiniest of shards. In addition, once something is found, everything about it must be properly recorded, including not only its location, but also its orientation, what material it was found in, and, when digging in layers, the exact depth at which it lies.

Louis, though amazingly adept at knowing where to search, was not the most careful of technicians. This got him into trouble later, when he couldn't properly document one of his published finds. The careless error was to dog him for the rest of his long professional career and might have finished off a lesser man.

His frequent use of newspapers to report his discoveries didn't help either, earning him the appellation of "Abominable Showman." *Punch,* not to be outdone when describing one of his finds, referred to "Oboyoboi Gorge." The journalistic accounts both annoyed and tantalized the academic community.

After repeatedly hearing these noises from Africa, the academic world didn't quite know what to make of them. At a scientific conference held in 1947, a diverse group of researchers in the field were finally convinced of our African origins by a reputable fossil expert, Robert Broom. Broom showed that australopithecines, including Dart's Taung child—dated to between 1 million and 2 million years ago—did indeed lie in the human line. By that time, academicians were becoming familiar with Leakey's name, and with that of Olduvai Gorge.

Olduvai Gorge

Louis, and then his wife Mary, made Olduvai Gorge their own. Mary, who learned her paleoanthropology from Louis, learned well, for many of their famous finds were actually made by her, although Louis often got the credit. She was actually the first of the fieldworkers to excavate in such a way that the necessary data could be accurately determined and recorded.

In the summer of 1959, after an unbelievable 30 years of stubborn, persistent searching, usually under extremely trying circumstances, they struck gold. Mary not only unearthed a remarkable skull, but she also found tools along with it. This was a real surprise, for toolmaking was thought to be another marker of advanced humanity. Saying that she found a skull makes the process sound a lot easier than it was: It came out in some 400 bits and pieces, which she had to fit together. It was like putting together a massive, three-dimensional jigsaw puzzle.

The result was a skull that showed huge jaws and a gorilla-like crest, yet which had certain human characteristics. It was,

Louis felt, worthy of its own species designation, and he christened it *Zinjanthropus boisei* (*Zinj,* an old name for East Africa; *anthro,* "man"; and *Boise,* the name of one of his financial supporters). Describing it as the earliest human ancestor, and, in fact, the missing link, he created a sensation. Later dated at 1.75 million years old, the skull added evidence that maybe Louis had been right all along. It was eventually reassigned to the *Australopithecus* group and is now known as *Australopithecus boisei.* That find alone changed the Leakeys' lives; suddenly, they were famous, and money started to come in far more easily, permitting them to mount better equipped, better supplied, and more fully staffed expeditions.

Richard Leakey

At the same time, Mary and Louis's son Richard was growing up. Independent like his father, he had no desire to bask in his parents' fame and, at first, he went off in a different direction. For a while, he put his naturalist skills to work by running a safari company.

In his wanderings, however, another region had caught his eye—Koobi Fora, Kenya, well north of Olduvai but still within the Great Rift Valley. Then, with even fewer credentials than his father—he never went to college—he assembled a team and came up with a variety of finds, but nothing of great moment.

In 1972, however, the last year of Louis's life, Richard presented him with a major find: a magnificent skull, one of the most complete ever dug up. The braincase was larger than that of earlier fossils, and without the prominent brow of previous finds. Louis beamed; the Hominid Gang had struck again. Also, here was solid evidence of what Louis had long believed—that a true ancestor in the human line, one with a big brain, had lived in Africa 2 or more million years ago. After some initial confusion about its age, it was dated at 1.9 million years old. Richard now had the distinction of having found the oldest member of the Homo clan, our own. In 1973, when he published the find, he assigned it to its own species, *Homo habilis,* the toolmaker.

Though Richard was clearly part of the Hominid Gang and had inherited some of the Leakey luck, he was in many ways very different from his father. A small example: Rather than give his find a pet name—his parents often referred affectionately to Zinj as Nutcracker Man, for its huge jaws, and even "Dear Boy," for obvious reasons—he simply called it "1470," its actual designation in the field. In fact, he referred to all his finds by their field numbers, hoping in this way to keep the inevitable discussions and disagreements on a less emotional level. It was a tendency that he tried to maintain in the controversy that was to come, though not always successfully.

He wrote later that 1470 "did for me what Zinj did for Louis; it made me famous, put me on the international stage."[3] This brought along with it some negative aspects. The London *Economist,* in its review of one of Richard's books, put it this way: "Depending on which side of the Atlantic you come from, Mr. Leakey is either a possessive and obstinate ignoramus with a talent for self-publicity, or the last great amateur scientist, who is right far more often than his better trained rivals in his guesses and interpretations of fossils."[4] Not that there was ever any question about his level of experience: Young Richard and his two brothers went out time after time with their parents and had more field experience under their little belts than other academics ever get in their entire careers.

Three of Richard's succeeding expeditions in the early 1970s were highly successful as well. So, in the space of 4 years in his own choice location, he had come up with as many important finds as his father and mother had in 30 years of painstaking research. By 1979, at age 35, Richard Leakey—the amateur— had become a scientific superstar. Even his mother, who had established herself as an important name in her own right, could not compete. Like Louis, Richard was a far better publicist than Mary, not only for himself but also for the wider world of paleoanthropology itself.

Like his father, Richard felt a drive to keep paleontology a hot topic, which of course helped considerably in the raising of funds. As was the case with Cope and Marsh a century earlier, the expeditions were expensive, and digging up funds was no easier, or less important, than digging up fossils. Fund-raising was a constant chore, and he had to charm the public, to gain their support, as well as those who distributed the funds.

This required a constant round of public appearances, opening exhibits, and lecturing here, there, and everywhere. He often joked about his lectures, saying that they were the only time he attended college. But they always drew a full house.

He was now hobnobbing with the rich and famous. He also left his father's foundation to establish his own. In this way, he not only raised large sums for research but also had a solid say in who was to get funds and where they could spend them.

In the meantime, in an astonishing repeat of his father's personal life, Richard had divorced his first wife and married a budding paleoanthropologist, Meave. Like Mary Leakey, Meave took to the field easily and well. The Hominid Gang was flourishing. As for Richard, he was king of the mountain.

Lucy Blazes Forth

A young American, Donald Johanson, later described where he had stood vis-à-vis the Leakeys at the time: "I was still in high school when I read about Zinj in the *National Geographic*. The name Olduvai, with its hollow, exotic sound, rang in my head like a struck gong. I was about to graduate, and despite what my mentor Paul Leser had been telling me about the virtues of chemistry as a profession, I began thinking more and more about anthropology. Leakey's experience was proof that a man could make a career out of digging up fossils.

"I went off to college," he continues, "and Leakey promptly jolted me again. In 1962 there came a report that he had found another hominid fossil at Olduvai, this time not an australopithecine but a true human [*Homo habilis*]." The shocker in the report "was the age of this new *Homo:* about 1.75 million years, the same age as Zinj. At one stroke Leakey and his associates had tripled the known age of humans."[5]

Johanson was obviously feeling the pull of the missing link even then. By 1970, while he was beginning work on his doctoral dissertation, events took an ironic turn. An acquaintance of Richard Leakey, geology student Maurice Taieb, had been piecing together the geological history of the remote deserts of Ethiopia. Taieb was particularly interested in a region known as the Afar triangle, which was actually the northern end of the Great Rift Valley. "People were just beginning to understand the plate tectonics theory," Taieb later recalled, "and so I thought I would study this area for my dissertation."[6]

Richard knew Taieb and, after seeing some fossil specimens Taieb had found, suggested that he take along a paleoanthropologist on future field trips. Richard recommended Johanson. Although Johanson had not yet completed his dissertation and was advised that he'd be wasting his time in the Afar, he decided to go. Taieb and Johanson obtained letters of support from Louis shortly before his death, which helped them get funding, and by 1973, they had set up camp in a bleak sun-bleached area of Afar, called Hadar.

Jon Kalb, another member of the Hadar group, recalls: "Johanson was obsessed with finding hominids. He wanted to monopolize the expedition, to make the search for hominids its only purpose."[7] Richard visited the camp to see how they were getting on and asked Johanson whether he really expected to find hominids there. "Older than yours," Johanson answered, then added, "I'll bet you a bottle of wine on that." "Done," said Richard.[8]

A year later, in the fall of 1974, Johanson scored. He and his team came up with almost 40 percent of a skeleton; though dated at over 3 million years old, and only 3½ feet tall, it was strikingly similar to our own. Nicknamed "Lucy," the find rocketed Johanson into the paleoanthropological firmament with a speed that exceeded even that of Richard's ascent.

What brought Richard charging out of his corner (figuratively) was not the discovery, but the interpretation put forth by Johanson and his colleagues, especially Tim White. White, an early admirer of Louis who had also worked for Richard, eventually fell out with him. White helped convince Johanson that Lucy represented a new species, which they named *Australopithecus afarensis*. They further maintained that Lucy's pelvis, femur (upper leg bone), and tibia (lower leg bone) showed that she was bipedal.

Johanson knew well that the term *oldest human* had a magical quality, and that's basically what he claimed, particularly after another successful field season in which his group came up with the remains of at least 13 other individuals, which he called the "First Family." What were his own feelings about finding Lucy? Was he concerned strictly with professional advancement or, perhaps, with advancing knowledge?

Virginia Morell, who wrote an extensive biography of the Leakeys, maintains that Johanson was obsessed with the Leakeys, especially Richard. In describing the magic moment of discovery, she writes, "As he held aloft leg, arm and hand bones for the camera, he called out, 'Hey Richard, look at this one! This one's a good one! I've got you now, Richard! I've got you now.'"[9] Morell also quotes Taieb as saying that after discovering Lucy, "Johanson begins to act as if he is the leader. He wants everything for himself, and it was all because he wanted to pass Richard."[10]

Action and Reaction

There are conflicting reports of Richard's reaction to the finds, but there is little doubt that he was not happy about what happened next. At a press conference called to explain their findings, Johanson's team trumpeted "an unparalleled breakthrough in the search for the origins of man's evolution. . . . We have in a matter of merely two days extended our knowledge of the genus *Homo* by nearly 1.5 million years. . . . All previous theories of the origins of the lineage which leads to modern man must now be totally revised."[11]

Johanson, this upstart in sheep's clothing, was turning the whole field upside down. Although there were still some doubts about the date for Richard's 1470 find, Johanson's claim that his team had pushed back the origins of humans by 1.5 million years was, Richard felt, a premature claim of victory in the antique derby.

In spite of this and the other claims, however, there was no real break in relations, and the two groups remained in touch.

In fact, at a meeting not long after, Richard and Mary brought along some of their finds from Olduvai and from Koobi Fora, hoping to get the dating problem straightened out. Johanson later reported in his book *Lucy* his feeling that Richard and Mary were doing everything they could to avoid facing up to the dating problem, but Richard felt that his group was just trying to figure out what was going on. Taieb, who was at the meeting, later reported that things were beginning to heat up, for Mary as well as for Richard.

Creating a new species is always a traumatic event. In this case, Johanson's introduction of *Australopithecus afarensis* created a storm on several fronts. The official announcement was to take place at a Nobel symposium held in 1978. Mary attended the symposium and was infuriated when Johanson announced that he was including a number of her fossil finds in his classification. Though she had probably known of his intention, the public announcement was particularly galling because his classification ran exactly counter to the position long held by the Leakeys.

The situation was exacerbated when Johanson later published his position, with co-author Tim White. In it, he set up Lucy and his "First Family" as the earliest group that could be considered true human ancestors. If he was right, here indeed was the, or at least a, missing link.

According to his classification, *A. afarensis* sits at the base of a neat Y-shaped tree. Lucy, the Mother of Mankind, forms the stem, which branches off in one direction to *Homo habilis,* which in turn leads eventually to *Homo sapiens,* modern man. The other branch of the Y leads to Louis Leakey's *A. boisei* and thence to extinction. This directly contradicted the Leakeys' belief that the human line began much earlier. Thus were several lifetimes of work put on the line, and with some Leakey fossils used as ammunition against their own position. Also, Johanson thereby claimed the title to being finder of *the* missing link.[12]

It's hard to judge just what Richard and Mary's true feelings were. Richard, as usual, claimed that he was really just looking for the truth. He's never expressly said so, but he had to be thinking that Johanson, in his anxiety to claim the title, was being premature. Leakey's position was that there was simply not enough fossil evidence to support Johanson's claim; that there was room for Lucy in existing categories; and, finally, that she could have been placed in a "suspense" account, as he had done with one of his own finds earlier in his career.

A strong argument against Johanson's inclusion of Mary's fossils was that he was lumping together specimens separated in time by half a million years and in space by a thousand miles. In a letter to a colleague, Mary called the Johanson team's work "slovenly," and she asked him to "join issue" against them.[13]

Things heated up even further when Johanson published his book on Lucy. Among his comments about Mary's reaction: "She hit us . . . with a hairsplitting obfuscation about nomenclature, about errors she claimed we had made in naming our new species."[14] Nevertheless, the Leakeys and Johanson were still speaking, which set the stage for what appears to be Johanson's and Richard Leakey's final personal confrontation.

If anyone still had doubts that paleoanthropology had entered the public domain, they were dispelled when Walter Cronkite invited both Johanson and Richard to appear on his influential and widely watched television program, *Universe*. According to Johanson, Leakey had been claiming that the rivalry between them was a myth and largely the invention of the press. "I thought that was misleading, and I welcomed the opportunity to meet with Richard on the record."[15] Richard, on the other hand, believes he fell into a trap, though not necessarily of Johanson's making, for he had been assured that this would not be a debate, but rather a discussion about creationism and human evolution!

It was not that Leakey feared debates, but he felt that because the fossils under discussion were Johanson's, he would be at a distinct disadvantage. Sure enough, Johanson had with him some props, including an *A. afarensis* skull. Further, says Johanson, "As soon as the cameras began to roll, it became apparent that a debate was just what Cronkite was looking for."[16]

Then Johanson presented his version of the human tree on a diagram and looked over at Leakey. Richard, angry that he had allowed himself to be trapped like this, drew a big X across it. Then, maintaining his usual position that we just don't have enough fossil evidence for a solid decision, he simply drew a big question mark on the other side. Later, Leakey described the show as "unfortunate." Johanson still insists, "I won!"[17]

That was back in 1981. They haven't spoken since,[18] but the harmful effects linger on. In 1984, Leakey began to withdraw from the field. Some think the feud had something to do with his withdrawal; he insists that he just had other interests and wanted to pursue them. Although he continued his administrative work with the National Museums of Kenya, and with the new Institute for Primate Research, an outgrowth of his father's institute, he did withdraw from the outside work. He shied away from the conferences and meetings that had been so important in his early days, particularly from any meeting where he might run into Don Johanson.

Leakey's name, however, was still high up in paleoanthropological circles. In 1984, for example, the American Museum of Natural History in New York sponsored a major exhibition and meeting called "Ancestors: Four Million Years of Humanity." The organizers wanted to show the originals from among the major fossil finds, including Dart's Taung child, Louis and Mary's Zinj, Richard's 1470, and Johanson's Lucy. Johanson was to give the keynote. Richard was also invited to speak and to show some of his fossils. He not only refused to participate, but also refused to lend any of the Leakeys' original fossil material, citing fears for their safety.

Mary, too, was invited and did appear. In her talk, she praised the organizers for the well-run event; but, echoing Richard, she also pointed out that these irreplaceable fossils had been gathered into a single room where a religious terrorist (undoubtedly referring to creationists) could conceivably demolish the entire legacy. That comment wasn't the end of it, however. Other museums also refused to lend their fossils. As had happened often in the past, Richard, whose power was fabled, was blamed. His power, however, actually did not extend beyond Kenya's borders, and so most of the complaints were unjustified.

Among Richard's other activities was a stint as director of the Kenya Wildlife Service from 1989 to 1994. Exercising a strong hand, he angered many people. In 1993, perhaps as a result of sabotage, and perhaps by pure accident, a plane he was riding in went down, and he lost both legs. Although he has shown great courage and has managed to carry on in many areas, fieldwork is very difficult for him. As a result, his wife and collaborator Meave has taken over many of the field functions in their expeditions. Richard has also entered politics.

What Do We Mean by "Human"?

One of the difficulties faced by paleoanthropologists is that there has not been a solid, consensually accepted definition of what we mean by "human." In fact, it is through the paleoanthropological finds that the definition has been evolving. The early idea was that (a) our humanness began when, for some reason, our forebears dropped down from the trees, leaving their hands free from brachiating to evolve into toolmaking devices, and (b) our brains started to enlarge at about the same time.

By the early 1980s, all the evidence showed that the expansion of the human brain took place no earlier than 2–3 million years ago. Bipedality, however, had been dated back to at least 4 million years. Among the evidence was a remarkable set of tracks unearthed in 1978 by Mary Leakey in Laetoli, Tanzania. To Mary's chagrin, Johanson included these tracks as part of the evidence for his *afarensis* claims.

There was also agreement by both sides that the two main hominid lines, *Australopithecus* and *Homo,* converged in an earlier ancestor. As we have already seen, Johanson said that Lucy and her clan belonged at the base of the split, and that the split occurred after her time, between 4 and 3 million years ago. Richard Leakey argued that Lucy was merely another, if earlier, australopithecine, and that the common ancestor would be found much earlier, as long as 7 or 8 million years ago. He also felt that the Y shape was just too simple, that the "tree" should more properly be thought of as a bush.

In addition, Leakey maintained that his argument was more in line with evolutionary thinking. In 1992, for instance, he pointed out that the *Alcelaphini,* a "tribe" of African antelope that includes the blesbok, hartebeest, and wildebeest, became extremely effective and successful grazing machines. The animals first appeared a little more than 5 million years ago, represented by one species. Living on tough forage, they spread over much of sub-Saharan Africa and now have 10 existing branches in their evolutionary bush.[19] "In shape," he writes, "the evolutionary history of the *Alcelaphini* tribe looks like a flat-topped acacia tree." He still believes that new finds will support his idea that a more complex evolutionary tree is the likely configuration for humans as well.

Evolutionary biologists have tended to agree, while anthropologists have leaned in Johanson's direction. Nevertheless, the decibel level dropped for a while.

New Finds

As genetic techniques for identifying and classifying species have improved, researchers' hopes have risen for dramatic improvement in our understanding. Some genetic evidence has been advanced, suggesting that the human and ape lines separated at least 5 million years ago, and maybe even as much as 7 million years ago, as Richard had hypothesized. But because the fossil record does not yet go back that far, the molecular genetic evidence cannot be confirmed.[20]

More recent evidence, however, has come in, which has heated up the argument once again. One of the major contributors has been the newest member of the Leakey clan: Meave. Meave was interested in yet another region in the Great Rift Valley, a desolate spot named Kanapoi, the sediments of which were known to date back to 4 or 5 million years. Her hunch paid off. In 1994, she unearthed hominid specimens that, along with some others found in the area, date back to between 4.2 and 3.9 million years ago. The excitement lies not only in the great age, but also in the fact that the specimens appear to be from a species different from any that had been found before.

The team has named it *Australopithecus anamensis*[21] (not to be confused with Johanson's *afarensis*).

At roughly the same time, Tim White (who has recently fallen out with Johanson) and his colleagues came up with even older bones at Aramis, Ethiopia. Dated to 4.4 million years ago, they seem to come from yet another species. While its name is in dispute, the important finding is that these various species have almost certainly overlapped.[22]

This overlap of species means that the simple linear idea, of one species eventually evolving into another in a simple straight line, goes out the window. Thus, the situation, rather than simplifying, is even messier than it was before. But one thing seems clear: the bushy description favored by both Louis and Richard seems more sound than ever.

Then, to confuse matters even further, two Swiss anthropologists who have studied Lucy's skeleton argue that she may be a "he"! While the argument is complex, it centers around the shape and size of the pelvis. They argue that Lucy's pelvis is simply not large enough to accommodate an australopithecine baby.

What this means, of course, is that—if they are right—Lucy is not a sexually dimorphic small female member of *afarensis,* but a full-size male member of a different species altogether. Martin Häusler, one of the researchers, states: "I cannot say for certain that Lucy was male. What I *can* say is that she did not belong to a species with great sexual dimorphism in body size."[23]

In any case, the controversy is wide open once again. Is the missing link still missing? Maybe, maybe not.

As president of the Institute of Human Origins in Berkeley, California, Johanson remains active in the field. Although he seems to have mellowed a bit, the fires have not gone out altogether. In his 1994 book, *Ancestors: In Search of Human Origins* (co-authored with Lenora Johanson and Edgar Blake), he was still writing as if his approach was the right one, ignoring such evidence as that of Russell Tuttle, an expert on comparative anatomy at the University of Chicago. Tuttle maintains that the footbones from the Hadar location are different from the humanlike footprints found by Mary Leakey in Laetoli.[24]

Further, in a 1996 *National Geographic* article, Johanson couldn't resist inserting a small dig: "She [Lucy] may not be our oldest ancestor, but she remains the best known."[25]

Richard Leakey has refrained from further comment; keeping true to his usual laid-back self, he is letting others fight the good fight.

CHAPTER 10

Derek Freeman versus Margaret Mead

***

Nature versus Nurture

A *Business Week* article about sociologist Sherry Turkle refers to her as "the Margaret Mead of cyberspace."[1] Even if you've never heard of Turkle, you have a pretty good idea that she's some sort of leading light in the computer field, that her ideas are provocative, that her writings and lectures are interesting and accessible, and that she has developed a following that extends beyond academe. This admiring use of Mead's name is appropriate. When she died in 1978, President Carter mourned her death and stated that she had "brought the humane insights of cultural anthropology to a public of millions."[2]

Mead was not only a world-renowned scientist, but also a guru to vast numbers of young people during the turbulent 1960s; an adviser to many parents through lectures and such writings as her column in *Redbook* magazine; and a consultant to governments on social policy. In her own field, she was a tireless worker, having studied and written about seven different South Seas cultures. By the end of her life, she had published over a thousand articles and two dozen books. A *New York Times* article said "she must be regarded as a pioneer whose innovations in research method have helped social anthropology come of age as a science."[3]

One of her innovations was writing in a way that the public could understand. Rather than filling the text with the more typical detailed observations that could be statistically organized, she relegated this material to appendixes at the end. Though her accessibility endeared her to the public, it irritated many of her more hidebound colleagues. Even more irritating was that her very first publication, *Coming of Age in Samoa*, catapulted her to celebrity status with astonishing speed and established her as a force to be reckoned with.

Nature versus Nurture

At the time of the book's publication (1928), the scholarly world was still enmeshed in a long-running debate about the origins of human behavior. A wide variety of scientists, scholars, and government workers had taken the rediscovery of Mendel's work on genetics and had built upon it a pseudoscientific edifice, which declared that human behavior is genetically determined. Unfortunately, this provided powerful ammunition for racists and eugenicists. The eugenicists sought to improve the human species through what they called "selective breeding."

On the other side stood the "nurturists," or cultural determinists, who argued that human behavior is largely or even entirely a result of culture and environment. Therefore, they argued, although selective breeding may work with animals, it is a useless and dangerous idea when applied to humans.

It seemed there was no meeting ground between the two groups. The inevitable result was intellectual chaos, with some geneticists even arguing that they had superseded Darwin. The argument turned nasty when eugenic thinking began to slide into virulent racism. Legislators and politicians were pounded with the idea that now, finally, we could do something about the problems of society. Among the solutions: sterilizing "inferior" individuals and restricting immigration from "less developed" societies. Although racism was found among some of the early anthropologists, a few important figures such as Franz Boas, Mead's mentor at Columbia University, argued strongly against it.

Mead, almost single-handedly, took the heart out of the eugenics movement—and she did it with that most unlikely of

weapons: a book that included some surprisingly romantic, almost flowery, prose. In one of the chapters, "A Day in Samoa,"
she wrote, "As the dawn begins to fall among the soft brown
roofs, and the slender palm trees stand out against a colour-
less, gleaming sea, lovers slip home from trysts beneath the
palm trees or in the shadow of beached canoes, that the light
may find each sleeper in his appointed place."[4] Also "at last
there is only the mellow thunder of the reef and the whisper of
lovers."[5] Trysts? Lovers? What does all this have to do with eugenics and racism?

Negative Instance

Albert Einstein once pointed out that no amount of experimentation could ever prove his theory of relativity right, but at any
time, a single reproducible experiment could prove him wrong.
Margaret Mead's *Coming of Age in Samoa* was the bombshell,
the single experiment that did for anthropology, sociology, and
psychology what Einstein feared in relativity. It blasted the
eugenics/naturist edifice to pieces, at least for a while.

Credit is usually given to an inspired insight that came out of
the combined thinking of Mead and Boas, her Ph.D. thesis adviser at Columbia University. But the idea really came out of
her broad range of interests, apparent even at the tender age of
23. These included a psychological orientation—she had been
working toward a master's degree in psychology before switch-
ing to anthropology; in fact, *Coming of Age* is subtitled *A Psychological Study of Primitive Youth for Western Civilization.*
Her humane instincts and identification with young people
surely played a part, as did her growing familiarity with a wide
variety of adolescents, including young immigrants entering the
United States.

All of these factors led her to think about adolescence in various societies. It seemed to her that there were more differences than similarities. Out of this observation came the brilliant idea of challenging the hereditarian position via what has
since come to be called a "negative instance."

If she could find a society in which adolescents do not go through the storm and stress that American young people seemed to suffer, then it would be clear that what Westerners called adolescent "turmoil," then believed to be a powerful natural behavior, was culturally produced. She also suspected that somewhere in the islands of the South Seas she might find a culture in which the passage from childhood to adulthood did not cause the same pain that it did in the Western world. And find it she did—in several villages in American Samoa. Simply stated, her conclusion was that the Samoan culture was able to provide a relatively smooth transition for the group of 50 young women she studied as they came of age.

Her book-length manuscript quickly found a commercial publisher. To make the book more marketable, her editor suggested that she should include a few additional chapters in a popular style and try to generalize some of the text so that it applied to our own culture.

In typical Mead fashion, she took the idea and ran with it, drawing bold comparisons between the Samoan and American cultures, and not necessarily in the latter's favor. For example, "it is not pleasant to realize that we have developed a form of family organization which often cripples the emotional life, and warps and confuses the growth of many individuals' power to consciously live their own lives."[6]

Academic critics wondered, Who was she to presume to tell us how to raise our kids? Another sticking point in many craws was her conclusion that the Samoans' ease of passage was at least partly due to a much freer atmosphere. She saw the Samoans as gentle, peaceful, and devoid of jealousy. Most important, however, she found that, with certain exceptions having to do with high social position, they also condoned a kind of adolescent free love. As a result, sex among their young people was a "natural, pleasurable thing,"[7] which helped smooth their transition from childhood to adulthood—that is, the coming-of-age years.

In contrast, among America's young, "when there is added to the pitfalls of experiment the suspicion that the experiment is wrong, and the need for secrecy, lying, fear, the strain is so great that frequent downfall is inevitable."[8] This obviously did not sit well with many of Mead's American readers, especially those raised in the strict authoritarian atmosphere permeating much of America.

Still, she had struck a powerful chord, and in addition to creating an adoring public, she also attracted a wide circle of admirers in the anthropological, sociological, and psychological fields. For 55 years, then, the situation remained essentially in stasis, with support continuing to flow her way. In 1972, the respected anthropologist E. Adamson Hoebel called *Coming of Age* a "classic example" of the use of fieldwork as the equivalent to the experimental laboratory.[9] The "negative instance" had done its work well.

By the time of Mead's death, in 1978, her reputation still seemed secure; *Coming of Age in Samoa* was probably the most widely read anthropology book ever published, with editions in 16 languages, and with millions of copies in circulation, many to undergrads who cut their anthropological teeth on it.

Inevitably, there were some doubters, particularly after 1975, when Edward O. Wilson introduced his pronaturist ideas in his book *Sociobiology: The New Synthesis*. Some anthropologists also felt that she tended to go too far in some of her conclusions and generalizations, and that she was more effective as a popularizer than as a scientist. In fact, as Lola Romanucci-Ross put it in 1983, although "Margaret Mead towered above many of her generation with her multitudinous talents. . . . She has never been accused of having been the most meticulous and persistent of linguists, historians or ethnographers."[10] Mead also traveled all over the place and made lots of money, which led to other muttering.

These voices of dissent, however, were quiet ones. Perhaps the grumblers didn't care to challenge Mighty Mom—or, as some claim, perhaps they didn't dare to. Although Mead was generous and helpful to those she liked, she could be imperious and impatient as well. One of her favorite words, "piffle," could be used to devastating effect. She also wielded considerable power over grant money and employment. So the complaints and grumbles were all small stuff—until. . . .

D-Day

On the morning of January 31, 1983, just four years after the
Johanson/Leakey article appeared, readers of the *New York
Times* were treated to a low-key headline at the lower left-hand
corner of page 1: "New Samoa Book Challenges Margaret
Mead's Conclusions." The first sentence reported, "a book maintaining that the anthropologist Margaret Mead seriously misrepresented the culture and character of Samoa has ignited heated
discussion within the behavioral sciences."

The new book was titled *Margaret Mead and Samoa: The
Making of an Anthropological Myth,* and the author was Derek
Freeman, Professor Emeritus at Australian National University,
who had spent many years studying the cultures of Western
Samoa. Again, the reader had to wonder, why was it placed on
page 1?

Well, perhaps it was because his book was being published
by the highly respected Harvard University Press. No, that can't
be it. Harvard's scholarly tomes rarely make the newspapers.
The real reasons for front-page coverage came later in the article, and this time they remind us of the Cope/Marsh newspaper
series, for Freeman was insisting that many of Mead's assertions about Samoa "are fundamentally in error and some of
them [are] preposterously false." Not only were Samoans not
prone to casual lovemaking, he maintained, but "the cult of virginity is probably carried to a greater extreme than in any
other culture known to anthropology." Just about everything in
Mead's book, he seemed to be saying, was in error. In a tele-
phone conversation with the *Times* reporter, he maintained,
"There isn't another example of such wholesale self-deception
in the history of the behavioral sciences."

Mead, unfortunately, was not around to defend herself, for
this was the sort of battle she would have enjoyed. Although
plenty of others leaped to her defense, they found themselves
in a strange predicament.

First, the *Times* article had appeared two months before the
official publication date of the book. Even more important,
Freeman had been brought halfway around the world for a series of interviews several months before the book's publication

date. His confident manner and willingness to take on all comers was appealing to talk-show hosts; his disparaging comments about cultural determinists didn't hurt either. The problem was that when responsible reporters got wind of the story, which didn't take long, and called other anthropologists for their reactions, these unhappy people had to comment before even seeing Freeman's book.

When the book finally came out, there was another media explosion, which of course was just fine with both the publisher and Freeman. Typically, this media attention would have died down after a while. Somehow, interest has not waned. Everyone, it seems, has had something to say—in books, reviews of the books, comments on the reviews, answers to the comments, and—of course—articles and papers, all showing an astonishing range of points of view.

Historians, sociologists, psychologists, and even psychiatrists have chimed in. Vera Rubin, director of the New York–based Research Institute for the Study of Man, published one of the strongest reviews in the the *American Journal of Orthopsychiatry.* She wrote that Freeman "pompously challenges the 'myth' of Mead's work," then describes his response as "presumably the behavioral science equivalent of exposing the Piltdown fraud. His methodology, however, is questionable at best; the conceptual orientation is parochial; and, upon careful perusal of his counterclaims, the acrimonious attack against Mead simply does not hold water."[11] Later, she added, "It is not unreasonable to reverse Freeman's accusations that Mead's book is based on mind-boggling contradictions."[12]

Tempers flared, with some curious results. The Northeastern Anthropological Association membership voted to censure Harvard University Press, as well as the *New York Times* (which published three articles on the dispute in its first week) and Freeman himself. The resolution was not adopted. The American Anthropological Association did, however, pass a resolution, noting with dismay the recommendation of Freeman's book as a holiday gift by the magazine *Science 83.*

Of course, it all depends on whose ox is being gored. Thomas Bargatzky, a German anthropologist, argues that Freeman's critique is not a personal attack on Mead, but that "Freeman was subjected to an amount of aspersions and vilification unprecedented in the history of anthropology."[13]

The Charges

One of Freeman's charges was that Mead was more interested in ideology (that is, in pushing the nurturist position) than in doing solid research, and that as a result she ignored all evidence against nurturism. As for her followers, he later wrote that "her account was received with something akin to rapture by the behavioristically oriented generation of the late 1920s."[14]

Her defenders countered with exactly the same charge, but of course turned around. Micaela di Leonardo, who teaches anthropology and women's studies at Northwestern University, zeroed in on "the rightist feeding frenzy surrounding Derek Freeman's 1983 attack on Margaret Mead's Samoa research."[15]

On the other hand, the inevitable response turned the argument around again. Freeman based his challenges on his years of studies in Western Samoa, and he has maintained that his conclusions—that Samoans differ from Mead's descriptions in many ways—are perfectly transferable to the American portion of Samoa. Not so, the Mead supporters argue, and they give many examples of how different Western Samoa is (larger, more highly populated, more developed, and so on)—not to mention the fact that Freeman didn't even begin his studies until years after Mead studied the Samoans.

In fact, Mead herself recognized a potential problem that could arise whenever descendants of her informants read the book, or if later workers tried to replicate or evaluate her work, a common occurrence in science. She even refused to update her book on the basis of later research. She wrote in the preface to the 1973 edition (the one being used here), "[The book] must remain, as all anthropological works must remain, exactly as it was written."

She also wrote in the same preface: "It seems more than ever necessary to stress, shout as loud as I can, this is about the Samoa and the United States of 1926–1928. Do not confuse yourselves and the Samoan people by expecting to find life in the Manu'an Islands of American Samoa as I found it. Remember that it is your grandparents and great-grandparents I am writing about when they were young and carefree in Samoa or plagued by our expectations from adolescents in the United States."[16]

Another of Freeman's claims was that he had "scientifically" refuted Mead's coming-of-age thesis. This claim has opened up a question that is often asked elsewhere: Can such soft disciplines as anthropology, sociology, and psychology really be classified as sciences? The responses range all over the board.

James E. Côté has entered the fray with a 1992 article and a follow-up book in 1994, both looking at the situation from the point of view of a sociologist, and one with a specific interest in adolescence. He argues that "the standards of proof in science place the onus on him [Freeman] to provide *irrefutable* evidence; if there are other plausible interpretations of the evidence he gives, then his conclusions cannot be considered any more definitive than Mead's, and the controversy becomes simply that of one interpretation against another. . . . Thus, innuendo, rumour, results of personal conversations, and quotations of material that are taken out of context or put together to form a 'creative collage' [all of which he charges Freeman with using] are all unacceptable."[17]

One example Côté gives of such "creative" thinking is Freeman's charge that both Boas and Mead were "absolute cultural determinists," meaning that they believed *all* behavior to be culturally determined. In contrast, Freeman claims, he is just trying to put some sense into the controversy, through his insistence on integrated anthropology, meaning one that includes both biology and society as cultural determinants. Mead's defenders point out, however, neither Boas nor Mead held the extreme position Freeman says they did.

Regarding this point, Marvin Harris, who was one of Mead's strongest critics prior to Freeman's book, adds: "the fact that major anthropology departments in the United States offer various courses in physical anthropology, primatology, medical anthropology, paleodemography, human biology, human genetics, and human paleontology (all of which have strong neo-Darwinian components) is largely because of Boas, not in spite of him."[18]

Lowell D. Holmes, who in 1954 replicated Mead's work (as well as could be done 28 years later), produced a book of his own, *Quest for the Real Samoa: The Mead/Freeman Controversy & Beyond.* Early on, he writes, "How it can be claimed that Franz Boas did not take the biological component into consideration in the study of human behavior is beyond comprehension."[19] Well, that's just what Freeman's defenders do claim, and they advance plenty of evidence to support the idea. There's even an ongoing battle over whether Boas ever accepted biological evolution.

Holmes's case is interesting. First, he has devoted half a century to the study of Samoan culture, so he is eminently qualified to comment. Second, he is by no means an unquestioning follower of Mead; he reports that his early relationship with her was very stormy, and that she gave his first book on his studies a "terrible" review.[20] Nevertheless, he maintains: "Although I differ from Mead on several issues, I would like to make it clear that, despite the greater possibilities of error in a pioneering scientific study, her tender age (twenty-three), and her inexperience, I find that the validity of her Samoan research was 'remarkably high.'"[21]

When Freeman learned of this conclusion, he questioned it, then wrote in a letter to Holmes (October 10, 1967): "You will also know, I take it, that Margaret Mead's name is execrated in Manu'a (as elsewhere in Samoa), for her writing. . . . Indeed, the people of Ta'u told me that if she ever dared return they would tie her up and throw her to the sharks."

"Let me say," Holmes continues, "that when Margaret Mead returned to Ta'u in 1971 to dedicate a power plant she was welcomed with open arms and showered with gifts and honors."[22] (Manu'a is a group of islands in American Samoa; Ta'u is one of these islands, and is the one containing the three villages in which she did most of her *Coming of Age* work.)

Irrefutable Direct Evidence

In 1991, Freeman wrote one of his many answers to charges that had been made against him, explaining why he felt that publication of his charges against Mead was fully warranted in 1983. Then he added, "Since then (cf. Freeman 1989), direct evidence has come to light—of a kind that could be presented in any court of law—that Mead was grossly hoaxed by her Samoan informants, and it is in the light of this and other direct evidence about her Samoan researches (cf. Freeman 1991) that *Coming of Age in Samoa* must *now* be assessed" (italics added).[23] The evidence is not spelled out, and the reader is referred back to the 1989 publication for details, but the impression of irrefutable "direct evidence" is left in the reader's mind.

Just how irrefutable is that evidence? In the 1989 article, Freeman begins, "In this brief communication, I report crucially important *new* evidence on Margaret Mead's Samoan researches of 1926" (italics in original). This "crucially important *new* evidence" involved Fa'apua'a, one of the young women Mead had interviewed many years earlier. Fa'apua'a had agreed at the time that sex on the island was quite free. More than 60 years later, however, she is saying just the opposite; further, she is claiming that she and Mead's other informants had been playing a joke on Mead. The 1987 interview with Fa'apua'a was filmed and became part of a documentary, *Margaret Mead and Samoa,* which was aired on TV in 1988 and then marketed to many anthropology departments across the country.

If Fa'apua'a's most recent statements are true, then of course Mead's work does fall apart. Is this evidence irrefutable? Although Freeman didn't actually say so, his reference to a court of law implied that his evidence would stand up in any court of law. Such a court implies a jury trial, however, and in such a trial, unanimous consent is mandatory. That is, all the jurors must agree.

That, clearly, has not happened, for there are plenty of demurrals from unanimous decision. If Fa'apua'a lied then, why are we to believe that she's telling the truth now? Are there any explanations for why she might be lying now? You bet.

The main one, advanced by a variety of Mead supporters, is that the situation has changed considerably since Mead did her work there. Samoa, as Mead took pains to point out, was a culture in transition. Even when she studied her Samoans, this society was by no means untouched. Missionaries had begun work there long before, and the society had been basically Christian for 80 years.[24]

Samoan society is complex, however, and old habits die hard, so some Samoanists have wondered whether the Samoans were Christianized, or whether their Christianity was Samoanized. Further, Fa'apua'a was at the time a *taupou,* one of the high-status maidens whose virginity was guarded most avidly. Since her time, the Christianization process has proceeded apace, along with a wide variety of other American influences, which are too complex to go into here. The result, however, may well be that Fa'apua'a and others from that time are now embarrassed by what she and Mead's other informants said then. She may feel that it is better, at this time, to call herself a reformed liar; perhaps she and her fellow informants can rewrite history.

Martin Orans, Emeritus Professor of Anthropology at the University of California at Riverside, who has gained access to and reviewed Mead's field notes, argues strongly that Mead was not taken in. In fact, he says, "not a single piece of information in any of the field materials is attributable to Fa'apua'a."[25] He also notes, "It is to Mead's everlasting credit that she preserved her field materials so that they may be examined. . . . Many anthropologists have confessed to me that they would never have had the courage to do so."[26]

In the meantime, the rhetoric continues to flow. In 1991, Freeman claimed that Mead was "cognitively deluded" during her study, and *Coming of Age* spread her delusion, producing one of "the more spectacular and instructive instances of collective cognitive delusion in the history of the human sciences."[27]

Among the questions raised by Mead's critics is one that I wondered about, too. With all that premarital sex going on, how come there weren't more pregnancies? Nicole J. Grant (Adjunct Lecturer in the Department of Sociology, Anthropology, and Philosophy at Northern Kentucky University) argues that the question is easily answered if we inquire more closely into what kind of sex Mead was describing. There are, Grant points out, other kinds of sex besides intercourse. "In traditional Samoa," she writes, "the most common word for sex was the word that means 'play.'"[28]

And so it goes: thrust and parry, charge and countercharge.

Articles and books, and even a play, continue to appear. The play, *Heretic*, is by David Williamson, a fellow Australian. In it, Mead appears in several guises, including Marilyn Monroe, Jackie Kennedy Onassis, and Barbra Streisand. Franz Boas, also the butt of powerful satire, appears in a salmon-color suit, red bow tie, and yellow and black boots. The only solid figure is Freeman, who appears as the heretic fighting against Mead's entrenched ideology. *Heretic* had a run in Sydney, Australia, (1996–1997) but has not been produced again. Freeman attended the play five times.

Five books on the controversy have appeared since 1990. This includes two by Freeman, who is still engaged in shooting down the Mead icon—and in defending himself against the many attacks on his earlier work. His latest, *The Fateful Hoaxing of Margaret Mead: A Historical Analysis of Her Samoan Research*, appeared toward the end of 1998. It covers much the same ground as his first book, but delves deeper into Mead's correspondence and field notes, and, especially, the testimony of Fa'apua'a.

Once again Freeman argues that he has presented "irrefutable direct evidence" that will surely convince readers of the correctness of his position. What's happened, of course, is that the book has simply prompted another round of battle. Martin Gardner, a well-known writer about what makes good science versus bad science, maintains that "Derek Freeman's conclusions are unshakable. Mead's . . . most famous book has become worthless."[29] Martin Orans, critiquing the book for the well-respected journal *Science*, feels that it "rides roughshod over the evidence."[30]

A double-barreled critique of Freeman's analysis of Mead's work appeared in the November/December 1998 issue of *Skeptical Inquirer*. James E. Côté points out that Freeman's story "is much simpler than the truth . . . which gives him an advantage over those scholars who attempt to fully represent Mead's research in Samoa"[31] (shades of the creationism controversy). Paul Shankman makes the point that "most professional anthropologists have lost interest in Freeman's argument. His critique of Mead and his history of the discipline are deeply flawed. Yet the same style of argument that has turned off anthropologists has attracted the media, and intelligent people have been drawn to Freeman's appeals to truth, science, and evolution." Shankman calls Freeman's work "an intellectual speedbump in the way of our understanding of Samoa, the work of Margaret Mead, and the state of anthropology today."[32]

Two Aspects

Is there to be no end, no resolution to the controversy? To help us find an answer, it's useful to divide the feud into two parts, the nature/nurture aspect and the Freeman/Mead aspect.

Regarding the nature/nurture debate, Lola Romanucci-Ross wrote in 1983, "I submit that Margaret Mead did prove her point that cultural beliefs determine behavior at least in the American society she so greatly influenced. Did we not, from the late 1930's to the late 1960's, move from socially repressed and denied sexual impulses to a socially approved sexual freedom?"[33]

Similarly, Côté points out that 1990s research supports Mead's position more than it does Freeman's.[34] Along the same lines, a fascinating study reported, in April 1997, in *Nature* magazine shows that "significantly more new neurons exist in the [brains] of mice exposed to an enriched environment compared with littermates housed in standard cages."[35] While this doesn't prove anything, it indicates just how important environmental influences are.

The truth, however, is that the nature/nurture skein remains as tangled as ever. Perhaps the Darwin of the twentieth—or twenty-first—century will actually tease out the relative importance of the two possible causes, if indeed it can be teased out. Perhaps these influences differ for each person. That would leave the situation approximately where it is now: a fertile ground for more papers.

As for a resolution of the personal feud, that's not much further along, either. We can, however, get some feeling about the consequences so far. Lowell Holmes says, "I am not sure whether the Mead/Freeman controversy has been a good thing or a bad thing for the science of anthropology."[36] On the other hand, he adds, "I must admit that I owe a great deal to Derek Freeman, as do most Samoan specialists these days, for rescuing all of our careers from obscurity."[37]

Côté also finds a sort of positive angle, suggesting that Freeman's critique has "alerted us to the limitations of Mead's research, and to potential problems with its generalizability. For this," he adds, "we are grateful."[38] Others also urge the profession to learn from the feud. Caton wonders "on what terms the methods and findings of the biological sciences, behavioral biology especially, can be integrated with anthropology and the social sciences."[39]

Orans, more critical, questions how such a "flawed work" as *Coming of Age* could have served as a stepping-stone to fame. He advances two main reasons: First, from the very beginnings of cultural anthropology, "its practice has been profoundly unscientific and positively cavalier in its willingness to accept generalizations without empirical substantiation. . . . It therefore often produces propositions that are untestable. Relationships and concepts tend to be so ill-defined that they provide too much 'wiggle room'—opportunity to claim that whatever test has been offered to falsify a claim has missed the intended meaning."[40]

Second, we, the general public, wanted Mead's findings to be correct; as he puts it, the fault "lies with those of us, like myself, who understand the requirements of science but failed to point out the deficiencies of Mead's work, and tacitly supported it. . . . Had the book [had] an opposite ideology, we no doubt would have ripped it apart for its scientific failings."[41]

On a more positive note, Professor Brad Shore of Emory University suggests that Freeman "brought out some of the contradictions and complexities of Samoan life."[42] More disturbing, perhaps, Côté points out that Freeman "inadvertently sounded the alarm of an unfolding tragedy in modern Samoa, namely the difficulties facing many young Samoans because of the cultural disenfranchisement brought on by Western influences."[43] He continues, "Moreover, the resolution to the controversy is not likely to be found 'somewhere in between' the two versions, as some have suggested. Rather, it appears to lie in part with [the need for] a complex and flexible culture that survives and maintains its integrity by appropriating the forces that seek to change it, and by defying easy definition and understanding. The protective strategy of incorporating an influence as much as possible before the influence incorporates the culture was noted by Mead."[44]

Thus, it appears that the controversy may have had some positive results. The obvious question remains, however: Could Freeman have done things differently? Did his critique have to be so strong? After all, criticism in science is not uncommon and is even expected. Further, it is not unknown for cultural anthropologists to come up with different interpretations of the same culture. The Tepoztlan peasants of Mexico were described very differently by Robert Redfield in 1930 and by Oscar Lewis 21 years later. In contrast with Freeman's slashing attack, however, Lewis admitted his debt to Redfield, even though he found much to criticize.[45]

So what was Freeman's motivation? Aside from his very obvious dislike of Mead's work, there is an even darker suggestion—namely, that he "was able to boost himself into prominence on the back of [Mead's] fame."[46] The sad truth may be that if Freeman had written a tamer, more dispassionate work, it might be lying quietly alongside another by R. A. Goodman, titled *Mead's Coming of Age in Samoa: A Dissenting View*— and published in the same year as Freeman's! How many have heard of it?

And how about Holmes, who knows the Samoan culture as well as anyone? His 1957 restudy of Mead's work remains an unpublished Ph.D. thesis, often studied by scholars, but not by the general public. In his 1983 review of Freeman's book, Holmes wrote, "I would have loved to play the giant killer, as Freeman is now trying to do. . . . But I couldn't. I had found the village and the behavior of its inhabitants to be much as Mead had described."[47]

Finally, there seems little doubt that if Freeman's objective was to sully Mead's reputation as a scientist, he has succeeded. As is most readily seen in the lives of politicians, when any person's life is put under a microscope, every flaw is magnified; rumors, exaggerated statements, and innuendo surface and stick in the reader's mind, whether valid or not. Mary Catherine Bateson, who is Mead's daughter but is by no means her automatic defender, feels that her mother's reputation has indeed been tarnished. She reports, "I still meet people who say, 'Oh, Margaret Mead! Wasn't her work entirely disproved?'"[48]

Bateson, who is Professor of Anthropology at George Mason University, also argues that "the whole process was wasteful and destructive and subversive of the possibility of using anthropological data responsibly in making societal decisions."[49] Of all the aspects to the controversy, this alone could cause Margaret Mead to turn over in her grave.

So, how do you get the world's attention? Mead did it in her way, Freeman in his.

EPILOGUE

The feuds included in this book showed a variety of ways in which resolution can take place. One method not included that I'd like to mention is resolution by a commission, or study group. This approach can be useful in helping resolve social issues, including such questions as the desirability of nuclear power or whether the greenhouse effect is really upon us.

Resolution of such issues is particularly important, for without it, society is hard put to make reasonable and widely acceptable decisions concerning what, if anything, to do about the problems inherent in such controversies.

One such vexing problem was solved in this way. The question was whether homosexuality is a disease.[1] For years there appeared study after study, paper after paper, angry response after angry accusation, with no resolution. Should it, for example, be included as a disease in the American Psychiatric Association's diagnostic manual of psychiatric disorders?

Finally, it was put to a vote among members of the association. Result: Members, by a vote of about two to one, decided it isn't a disease.

NOTES

Introduction

1. Personal communication, July 25, 1997.
2. Provine, 1988, pp. 27–29.

Chapter 1: Urban VIII versus Galileo

1. Galileo Galilei, 1632 (de Santillana translation), p. 44, footnote.
2. In Jacob Bronowski, *Ascent of Man*, 1974, p. 209.
3. Eurich, 1967, p. 185.
4. Galilei, 1632 (de Santillana translation), p. xv.
5. Ibid., p. xiv.
6. *Quarterly Review*, 1878, pp. 111–128.
7. Ibid., p. 120.
8. De Santillana, 1955, p. 9.
9. Galilei, in Drake, 1957, p. 144.
10. Letter to the Grand Duchess Christina, in Drake, 1957, p. 147.
11. In Drake, 1957, p. 154.
12. Drake, 1957, pp. 163, 164.
13. Galilei, 1632 (de Santillana translation) p. 70.
14. Ibid., pp. 121–122.
15. *Palatine Anthology*, Vol. IX, p. 577. In Higham, Thomas F., and Bowra, C. M., *The Oxford Book of Greek Verses in Translation* (Oxford, England: Clarendon Press, 1938), p. 643.
16. Galilei, 1632 (de Santillana translation), p. 468.
17. Ibid., p. 469.
18. Although Earth's tides are not caused by the planet's motion, the tides on a rotating Earth are different from what they would be on a stationary Earth. See, e.g., Burstyn, 1962.
19. Ibid., pp. 471, 472.
20. The relationship is strictly true only for short swings, but this in no way detracts from Galileo's brilliant observation. For further details, see Landes, David S., *Revolution in Time: Clocks and the Making of the Modern World* (Cambridge, MA: Harvard University Press, 1983).
21. Bailey, 1990, e.g., Chapters 1 and 2.

22. Redondi, 1987, e.g., pp. 323ff.

23. De Santillana, 1955, p. 2.

24. Personal communication, July 23, 1996.

Chapter 2: Wallis versus Hobbes

1. In Dick, 1949/1957, p. 147.

2. Ibid., p. 149.

3. In Skinner, 1996, v43n6, pp. 58–61, online, unpaged.

4. Ibid., pp. 58–61.

5. Quoted in Hinnant, 1977, p. 17.

6. In Dick, 1949/1957, p. 150.

7. Yukawa, Hideki. "Physics: A View of the Japanese Milieu," *Science,* May 20, 1983, p. 822.

8. In Dick, 1949/1957, p. 151.

9. See, e.g., Watkins, 1965, p. 16.

10. Hobbes, *Human Nature, or the Fundamental Elements of Policy,* in Molesworth (ed.), Vol. 4 (1840), p. 73.

11. Hobbes, *Decameron Physiologicum,* in Molesworth (ed.), Vol. 7 (1845), p. 129.

12. In Watkins, 1965, p. 17.

13. Hobbes, 1986 (1651), p. 186.

14. Mintz, 1962, p. 10.

15. Shapin and Schaffer, 1985, p. 319.

16. Mintz, 1962, p. 24.

17. Hobbes, 1986 (1651), p. 105.

18. See, e.g., Mintz, 1962, pp. vii and 55; and Mintz, 1972, p. 449.

19. Mintz, 1962, p. 22.

20. Scott, J. F. "The Reverend John Wallis, F.R.S.," *Notes and Records, Royal Society of London (1960),* Vol. 60, p. 57. (For a different view, namely, that the statement was "merely a conventional response" to some compliments from Hooke, see McClain, John W. "On the Shoulders of Giants," *American Journal of Physics,* June 1965, v33n6, p. 513.)

21. Eliot, 1910, p. 155.

22. Cohen, 1939, pp. 530, 531.

23. In Hazard, 1990, p. 307.

24. In Smith, Vol. 1, 1957 (1930), p. 204.

25. In Molesworth (ed.), Vol. 7, 1839–1845, pp. 187, 256.

26. Ibid., p. 316.

27. Ibid., p. 356.

28. My thanks to Professor Mintz for his help in decoding these barbs.

29. In Robertson, 1886, p. 179.

30. Ibid., p. 183.

31. Wallis, "Animadversions . . . No. 16, p. 289," *Philosophical Transactions of the Royal Society* (August 6, 1666). From the 18-volume, 1809 abridgement of the years 1665–1800, Vol. 1, p. 108.

32. Ibid., p. 110.

33. In Skinner, 1996, pp. 58–61.

34. Ibid.

35. Hobbes, 1986 (1651), pp. 111, 115.

36. Mintz, 1952, p. 99.

37. Gardner, 1960, p. 156.

38. Boyer, 1959, p. 178.

Chapter 3: Newton versus Leibniz

1. Merton, in Price, 1963, p. 68.

2. Westfall, 1980, p. ix.

3. Smith, 1934, p. 44.

4. Huxley, in Spitz, 1952, p. 343.

5. Frederick II, in Spitz, 1952, p. 341.

6. Merz, 1884, p. 126.

7. Bishop Atterbury, in More, 1962 (1934), p. 127.

8. In Hall and Tilling, 1977, v7, pp. xliv–xlv.

9. In Hall, 1980, p. 250.

10. In Watkins, 1965, p. 123. (Listed in Chapter 2.)

11. Westfall, 1980, p. 114.

12. Ibid., p. 174.

13. "On a Deeply Hidden Geometry and the Analysis of Indivisibles and Infinities," *Acta Eruditorum* v5, 1686; reprinted in Leibniz, *Mathematische Schriften*, Abtheilung 2, Band III, pp. 226–235.

14. In Westfall, 1980, p. 721.

15. In More, 1962 (1934), p. 398.

16. Letter to Thomas Burnet. In Hall, 1981, p. 95.

17. Hathaway, 1920, p. 167.

18. In Merz, 1884, p. 89.

19. Manuel, 1968, p. 971.

20. Ibid., p. 972.

21. A portrait alleged to be of the "Scientist Hooke" was published in *Time* magazine for July 3, 1939, p. 39. Very quickly, it was shown to be

spurious by M. F. Ashley Montague: "A Spurious Portrait of Robert Hooke (1635–1703)," *Isis,* v33, 1941, pp. 15, 16.

22. In Hall, 1980, p. 145.

23. See, e.g., Westfall, 1980, pp. 721, 722, and Hall, 1980, pp. 168, 177.

24. Newton, "Account of the Commercium Epistolicum," in Hall, 1980, p. 221.

25. In Hall, 1980, p. 221.

26. In Gillespie, *DSB,* Vol. 10, 1974, pp. 42–103.

27. Merz, 1884, pp. 196–197.

28. In More, 1962 (1934), p. 382.

29. In Westfall, 1980, p. 534.

30. See, e.g., Broad, 1981.

31. Newton, "Account . . . ," in Hall, 1980, p. 224.

32. Smith, 1934, p. 517, listed in Chapter 2 references.

33. Ibid., p. 166.

34. In Merz, 1884, p. 126.

35. Ibid., p. 189.

36. In Bury, 1960 (1932), p. 77.

Chapter 4: Voltaire versus Needham

1. "Victor Hugo's Oration," delivered at the one hundredth anniversary of Voltaire's death, May 30, 1878; in Besterman (ed.), Vol. 1, 1975 (1969), p. 52.

2. Letter to Jean Le Rond D'Alembert, June 26, 1766; in Brooks, 1973, p. 264.

3. Letter to Mme. Denis, Berlin, December 18, 1752; in Redman, 1949, pp. 487–488.

4. In Smith, 1957 (1934), Vol. 2, p. 132 (listed in Chapter 2); and in Orieux, 1979, p. 261.

5. Voltaire, 1752 (Fleming, 1901), *Works,* Vol. 19, Part 1, pp. 194, 196.

6. In Westbrook, 1972, p. 4.

7. In Meyer, 1939, p. 80; Joseph Needham's translation is slightly different: "one of the greatest triumphs of rational over sensual conviction," Needham, 1959, pp. 213–214.

8. *Philosophical Transactions,* n490 (1748), pp. 615–666; in Westbrook, 1972, p. 58.

9. In Westbrook, 1972, p. 28.

10. Ibid.

11. Ibid., p. 36.

12. Ibid., pp. 21–22.

13. Ibid., p. 108.

14. Ibid., pp. 21–23.

15. Ibid., p. 155.

16. Ibid., p. 107.

17. Ibid., p. 109.

18. Ibid., p. 181.

19. Meyer, 1939, p. 71.

20. In Joseph Needham, 1959, p. 218.

21. Gillespie, 1976, p. 85.

22. Westbrook, 1972, p. 86.

23. In Besterman, 1975 (1969), p. 550.

24. In Meyer, 1939, p. 60.

Chapter 5: Darwin's Bulldog versus Soapy Sam

1. In Desmond and Moore, 1991, p. 322; original in Napier, M., *Selection from the Correspondence of the Late Macvey Napier* (New York: Macmillan, 1879).

2. Darwin, 1859, pp. 63–64.

3. In Clark, 1984, p. 137.

4. Mayr, 1991, p. 99. See also Gould, 1995, and Eldridge, 1995.

5. In Clark, 1984, p. 125.

6. In Desmond and Moore, 1991, p. 488.

7. Ibid., pp. 488, 489.

8. Huxley, in Francis Darwin (ed.), 1958 (1892), p. 253.

9. Letter from T. H. Huxley to Francis Darwin, June 7, 1861, in Francis Darwin (ed.), 1958 (1892), p. 254.

10. In de Camp and de Camp, 1972, p. 159.

11. Huxley, in Francis Darwin (ed.), 1958 (1892), p. 252.

12. In Clark, 1984, p. 144.

13. Darwin (1859), p. 373.

14. In Clark, 1984, p. 145.

15. Mayr, 1991, p. 25.

16. Caudill, 1994, online (database: UMI Research 1), unpaged.

17. "Monkeyana," *Punch*, May 18, 1861, Imperial College 79:4; quoted in Caudill, 1994, online, unpaged.

18. This and the following descriptions of pamphlets and cartoons are all from Caudill, 1994.

19. Darwin, 1872 (1859), p. 357.

20. Dennett, 1997, p. 41.

21. Mayr, 1991, p. 128.

22. See, e.g., Bishop, 1996.

23. Numbers, 1992, p. 40.

24. Ibid., p. 41.

25. In Tierney, 1979, p. 361 (from Mencken's *Heathen Days 1943,* reproduced in Cairns, Huntington, [ed.], *H. L. Mencken: The American Scene—A Reader* [New York: Alfred A. Knopf, 1965]).

26. Clark, 1984, p. 281.

27. Ibid., p. 282.

28. Ibid.

29. Ibid., p. 283.

30. Ibid., p. 284.

31. Mayr, 1991, p. 132.

32. The First Amendment prohibits the promotion of any religious doctrine by the federal government; the Fourteenth Amendment extends the application of the First Amendment to the states.

33. For a visit to what might be called "creation science's headquarters," the Creation Science Institute, see Hitt, 1996.

34. See, for example, "Life at the Edge of Chaos," John Maynard Smith's review of *Darwinism Evolving: Systems Dynamics and the Genealogy of Natural Selection* by David J. Depew and Bruce H. Weber (Cambridge, MA: MIT Press, 1994). In *New York Times Book Review,* March 2, 1995, v42n4, pp. 28–30. (Are Depew and Weber denying or supporting evolution?)

35. See, for example, Horgan, August 1995, in which Horgan questions whether Stephen Jay Gould is really a Darwinian; also David Sloan Wilson's brief for group selection in the face of strong opposition, in Berreby, 1996; also Lewin, 1996.

36. Quoted in "Eugenia Scott Replies," *The Sciences,* March/April, 1996, p. 47.

37. A good example of the "irreducibly complex" argument can be found in Michael J. Behe's *Darwin's Black Box,* starting on p. 39. Dawkins's treatment appears in *Climbing Mount Improbable,* 1996, pp. 138–197.

38. Berlinski, June 1996.

39. Berlinski, 1995.

40. Quoted in a letter from John M. Levy; in *Commentary,* September 1996 ("Staff" in the bibliography), p. 15.

41. In Staff, September 1996, p. 30.

42. Darwin (1859), p. 255.

43. Dennett, in Staff, *Commentary,* September 1996, p. 6.

44. Dawkins, 1996, p. 75.

45. Herschel, in Francis Darwin (ed.), 1958 (1892), p. 232.

46. Dawkins, 1996, pp. 75–77.
47. Letter from John C. Frandsen, Chair, Committee on Science and Public Policy, Alabama Academy of Science, in *Scientific American,* December 1995, p. 10.
48. Colson, 1996, p. 64.
49. Numbers, 1995, online, unpaged.
50. Hammond and Margulis, 1981, p. 55.
51. See, e.g., Nesse and Williams, 1996 (1995).
52. See, e.g., Murdoch, 1996.
53. See, e.g., Wright, 1994.
54. See, e.g., Farber, 1994; also the review by Degler, 1996.
55. See, e.g., Wilson, 1996; also brief magazine pieces: Berreby, 1997, and Malik, 1996.
56. See, e.g., Lewin, 1997.
57. Grady, 1996, p. 81.
58. Ryan, 1997, pp. 8, 9.
59. Livingstone, 1987, p. 1; recent entries include Crook, 1994, and Ramsay's review of it, 1996.

Chapter 6: Lord Kelvin versus Geologists and Biologists

1. Casson, c. 1927, p. 42.
2. Ibid., p. 47.
3. Gillespie, *DSB,* Vol. 13, 1970–1980, p. 387.
4. In Casson, c. 1927, pp. 57–58.
5. In Smith and Wise, 1989, p. 167.
6. Ibid., p. 127.
7. Ibid., p. 525.
8. Ibid., p. 639.
9. Thomson's 1871 BAAS Presidential Address, in Basalla, 1970, p. 125.
10. Ibid., p. 126.
11. In Smith and Wise, 1989, p. 640.
12. Ibid., p. 642.
13. Ibid.
14. In Casson, c. 1927, p. 77.
15. Cowen, 1996, pp. 204–205.
16. In Smith and Wise, 1989, p. 642.
17. In Burchfield, 1975, p. 84.
18. Huxley, 1876, p. 249.
19. In Smith and Wise, 1989, p. 536.
20. Ibid., p. 603.

21. Twain, 1962 (1938), p. 212.
22. In Broad, 1996, p. C8.

Chapter 7: Cope versus Marsh

1. "Modern Light Literature—Science," anonymous, in *Blackwood's Edinburgh Magazine,* American Edition (August 1855), Vol. 41, p. 226 (bound volume).
2. See, e.g., Morell, 1997, pp. 36–45, and Padian, 1997, pp. 178–180.
3. Quoted in Schuchert and LeVene, 1940, p. 38.
4. Quoted in Lanham, 1973, p. 49.
5. Quoted in Ostrom and McIntosh, 1966, p. 14.
6. Schuchert and LeVene, 1940, p. 354.
7. Quoted in Shor, 1974, p. 46.
8. See, e.g., Colbert, 1995, p. 180. For opposing views, you'll have to go to the Internet. Try the subject *Rioarribasaurus* as well as *Coelophysis.*
9. Ibid., p. 18.
10. Colbert, 1968, p. 73.
11. From the *Herald,* January 13, 1890; quoted in Shor, 1974, p. 119.
12. Wheeler, 1960, p. 1171.
13. Quoted in Lanham, 1973, p. 118.
14. Ibid.
15. Marsh, O. C. "Introduction and Succession of Vertebrate Life in America" (address delivered before the American Association for the Advancement of Science, at Nashville, TN, August 30, 1877). In *Popular Science Monthly,* April 1878, v12, p. 697.
16. Quoted in Lanham, 1973, p. 121.
17. Quoted in Ostrom and McIntosh, 1966, p. 9.
18. Ibid.
19. Quoted in Spalding, 1993, p. 122.
20. Baur, *Herald,* January 12, 1890; quoted in Shor, 1974, p. 109.
21. Marsh, *Herald,* January 19, 1890; quoted in Shor, 1974, p. 169.
22. Quoted in Shor, 1974, p. 217.
23. J. B. Hatcher, "Osteology of *Haplocanthus,* with description of a new species, and remarks on the probable habits of the Sauropoda and the age and origin of the *Atlantosaurus* beds," *Memoirs of the Carnegie Museum,* Vol. 2 (1903), pp. 1–72; quoted in Lanham, 1973, p. 184.
24. Lanham, 1973, p. 269.
25. Quoted in Shor, 1974, p. 146.
26. Colbert, 1968, p. 146.

27. "Modern Light Literature—Science," anonymous, *Blackwood's Edinburgh Magazine*, American Edition (August 1855), Vol. 41, p. 226 (bound volume).

28. Sagan, quoted in "Carl Sagan, an Astronomer Who Excelled at Popularizing Science, Is Dead at 62," *New York Times*, December 21, 1996, p. A26.

29. Spalding, 1993, p. 152.

Chapter 8: Wegener versus Everybody

1. Wegener, 1966 (1915); original title, *Die Entstehung der Kontinente und Ozeane;* all references to the book are based on the 1966 Dover reprint of the fourth revised edition, translated into English and published in 1929.

2. Wegener, 1966 (1929), p. viii.

3. Ibid.

4. Ibid., pp. 2, 3.

5. Romm, 1994, pp. 407–408; see also Cowen, 1994, p. 110.

6. Wegener, 1966 (1929), p. 167.

7. Ibid., p. 1.

8. In Le Grand, 1988, p. 1.

9. Wegener, 1966 (1929), p. 16.

10. Ibid., p. 17.

11. In Hallam, 1983, p. 122. (Listed in "General Background.")

12. Ibid., p. 122.

13. In Sullivan, 1991, p. 15.

14. Ibid.

15. In Hallam, 1983, p. 136.

16. In Le Grand, 1988, p. 118.

17. Gohau, 1990, p. 196.

18. In Hallam, 1983, p. 124.

19. Ibid., p. 129.

20. Ibid., p. 135.

21. Quoted in Hallam, 1983, p. 136; original in Gevers, T. W., *Transactions of the Geological Society of South Africa* (1950), v52 (suppl), p. 1.

22. Including Frederick J. Vine and Drummond Matthews of Cambridge University in England, and Maurice Ewing and Walter Pitman at what is now Lamont-Doherty Geological Observatory in Palisades, New York.

23. In Hallam, 1983, p. 141.

24. Kerr, 1995a, pp. 1214–1215.

25. Nelson, quoted in "International Research Team Discovers Unsuspected Molten Layer in Himalayan Crust," online summary issued by Syracuse University, December 6, 1996.

26. Monastersky, 1996a, p. 356; also Nelson, 1996, pp. 1684–1687.

27. Monastersky, 1996b, p. 213; Pool, 1996, discusses work on the question, using simulation studies.

28. Staff, 1995, p. 123.

Chapter 9: Johanson versus the Leakeys

1. In Morell, 1995, p. 157.
2. In McAuliffe, 1994, p. 83.
3. Leakey and Lewin, 1992, p. 112.
4. *Economist,* November 21, 1992, p. 103.
5. Johanson and Edey, 1981, p. 98.
6. In Morell, 1995, p. 461.
7. Ibid., p. 463.
8. Ibid., p. 422.
9. Morell, 1995, p. 468, quoting "an eyewitness source."
10. Ibid., p. 467.
11. Ibid., p. 464.
12. In Leakey and Lewin, 1992, p. 346.
13. Morell, 1995, p. 492.
14. Johanson and Edey, 1981, p. 301.
15. Johanson and Shreeve, 1989, p. 89.
16. Ibid., p. 119.
17. Leakey, interview with Roger Lewin, November 19, 1985, in Lewin, 1987, p. 18.
18. McAuliffe, 1994, p. 39.
19. Leakey and Lewin, 1992, p. 109.
20. For further details on the two approaches, see Lewin, 1998.
21. Lewin, 1995, p. 14.
22. Ibid.
23. In Shreeve, 1995, p. 1298.
24. In Falk, 1995, pp. 108–110.
25. Johanson, 1996, p. 117.

Chapter 10: Derek Freeman versus Margaret Mead

1. Judge, Paul C. "Is the Net Redefining Our Identity?" *Business Week,* May 12, 1997, p. 100.
2. In Rensberger, 1978, p. 1.

3. Ibid.
4. Mead, 1973, p. 8.
5. Ibid., p. 11.
6. Ibid., p. 119.
7. Ibid., p. 112.
8. Ibid., p. 135.
9. In Hoebel's textbook *Anthropology: The Study of Man,* fourth edition (New York: McGraw-Hill, 1972), p. 8.
10. In Caton, 1990, p. 130; Romanucci-Ross, an anthropologist, is now in the Department of Family and Preventive Medicine at the University of California, San Diego.
11. Rubin, 1983, p. 550.
12. Ibid., p. 554.
13. From a review in *Pacific Studies,* v11 (1988), pp. 131–151; quoted in Caton, 1990, p. 256.
14. Freeman, 1983, p. 201.
15. di Leonardo, 1996, pp. 25–29.
16. Mead, 1973, pp. ix, x.
17. Côté, 1992, pp. 509, 510.
18. Harris, 1983, p. x; in Caton, 1990, p. 236.
19. Holmes, 1987, p. 2.
20. Ibid., p. viii.
21. Ibid., p. 103.
22. Ibid., p. 189.
23. Freeman, 1991, p. 327.
24. Côté, 1994, p. 77.
25. Orans, 1996, p. 92.
26. Ibid., pp. 18, 19.
27. Freeman, 1991, pp. 118–119.
28. Grant, 1995, p. 681.
29. Freeman, 1998, back cover.
30. Martin, 1999, pp. 164–165.
31. Côté, 1998, p. 30.
32. Shankman, 1998, p. 39.
33. In Caton, 1990, p. 129.
34. Côté, 1994, p. 18.
35. Kemperman, 1997, pp. 493–495.
36. Holmes, 1987, p. 175.
37. Ibid., p. ix.
38. Côté, 1994, p. 64.
39. Caton, 1990, p. 1.
40. Orans, 1996, pp. 124, 125.

41. Ibid., p. 12.

42. Shore, Report on a Symposium, "Margaret Mead and Anthropology: An Evaluation," in *Barnard Bulletin,* April 13, 1987; quoted in Caton, 1990, p. 285.

43. Côté, 1994, p. 64.

44. Ibid., pp. 10, 11.

45. See, e.g., Rensberger, 1983, p. 35. Also, personal interview with Morton Klass, Professor Emeritus of Anthropology at Barnard and Columbia.

46. In Harris, 1983, p. 18.

47. Holmes, 1983, p. 15.

48. In Monaghan, 1989, p. A6.

49. Ibid.

Epilogue

1. Caplan, 1988, pp. 22–23 (listed in Chapter 5 bibliographical material).

BIBLIOGRAPHY

I list here only the works that have been of use in the preparation of this book. Except for the first group, listed as general background, the list is divided into sections that correspond with the chapters.

General Background

Asimov, Isaac. *Asimov's Biographical Encyclopedia of Science & Technology.* Garden City, NY: Doubleday & Co., 1972.

Bolton, Sarah K. *Famous Men of Science.* New York: T. Y. Crowell Company, 1946.

Boorstin, Daniel. *The Discoverers.* New York: Random House, 1983.

Burstyn, Harold L. "Galileo's Attempt to Prove that the Earth Moves," *Isis,* 1962, v55, part 2, pp. 161–185.

Butterfield, Herbert. *The Origins of Modern Science, 1300–1800,* revised paperback edition. New York: The Free Press, 1965, 1957.

Engelhardt, H. Tristram, and Caplan, Arthur, editors. *Scientific Controversies.* Cambridge, England: Cambridge University Press, 1987. (Four case studies in a sociological treatise that seeks to develop a theory of how scientific controversies are resolved. Covers laetrile, homosexuality, safety standards, and nuclear power.)

Gillispie, Charles C., editor. *Dictionary of Scientific Biography (DSB),* 16 volumes. New York: Scribner, 1970–1980.

Hallam, A. *Great Geological Controversies.* Oxford, England: Oxford University Press, 1983. (Neptunists, vulcanists, and plutonists; catastrophists and uniformitarians; the Ice Age; the age of Earth; continental drift.)

Holton, Gerald. *Einstein, History and Other Passions: The Rebellion Against Science at the End of the Twentieth Century.* New York: Addison-Wesley, 1996.

Merton, Robert K. "Priorities in Science," *The Sociology of Science.* Chicago: University of Chicago Press, 1973.

Milton, Joyce. *Controversy: Science in Conflict.* New York: Julian Messner, 1980.

Officer, Charles, and Page, Jake. *Tales of the Earth. Paroxysms and Perturbations of the Blue Planet.* New York: Oxford University Press, 1993.

Raup, David M. *The Nemesis Affair.* New York: W. W. Norton, 1986 (paper, 1987). (Catastrophism.)

Shapin, Steven. *The Scientific Revolution.* Chicago: University of Chicago Press, 1996. (Especially sections of Chapter 3, on natural philosophy and its pre-Darwinian relationship to religion.)

Taton, R. *Reason and Chance in Scientific Discovery.* New York: Philosophical Library, 1957.

Williams, Trevor I., editor. *A Biographical Dictionary of Scientists.* New York: Wiley-Interscience, 1969.

Introduction

Provine, William. "Evolution and the Foundation of Ethics." *MBL Science,* Winter 1988, v3n1, pp. 25–29. (Marine Biological Laboratory, Woods Hole, MA.)

1. Urban VIII versus Galileo

Bailey, George. *Galileo's Children: Science, Sakharov, and the Power of the State.* New York: Arcade Publishing, 1990.

Biagioli, Mario. *Galileo Courtier: The Practice of Science in the Culture of Absolutism.* Chicago: University of Chicago Press, 1993.

Bronowski, Jacob. *The Ascent of Man.* Boston: Little, Brown, 1974.

De Santillana, Giorgio. *The Crime of Galileo.* Chicago: University of Chicago Press, 1955.

Dickson, David. "Was Galileo Saved by Plea Bargain?" *Science,* August 8, 1986, pp. 613, 614.

Drake, Stillman. *Discoveries and Opinions of Galileo* (translated, with an introduction and notes by Drake). New York: Doubleday, 1957.

Drake, Stillman. *Galileo at Work: His Scientific Biography.* Chicago: University of Chicago Press, 1978.

Eurich, Nell. *Science in Utopia: A Mighty Design.* Cambridge, MA: Harvard University Press, 1967.

Finocchiaro, Maurice A. *The Galileo Affair: A Documentary History.* New York: Notable Trials Library; Gryphon Editions, 1991. (Many of the relevant documents, translated into English, with an introduction by Alan M. Dershowitz.)

Galilei, Galileo. *Dialogue on the Great World Systems* (de Santillana translation). Chicago: University of Chicago Press, 1953 (1632).

Galilei, Galileo. *Dialogues Concerning Two New Sciences.* New York: McGraw-Hill, 1963 (paperback) (1638).

(Also in abridged text edition, by de Santillana, translation by T. Salusbury. Chicago: University of Chicago Press, 1955.)

Harsanyi, Zsolt de. *The Star-Gazer.* New York: G. P. Putnam's Sons, 1939. (Fictional account of Galileo's life, translated from Hungarian.)

Hummel, Charles E. *The Galileo Connection; Resolving Conflicts between Science and the Bible.* Downers Grove, IL: InterVarsity Press, 1986.

Koestler, Arthur. *The Sleepwalkers.* New York: The Universal Library (Grosset & Dunlap), 1963 (paperback) (original, Macmillan, 1959). (A readable history of the great astronomers, from Ptolemy to Newton.)

Kuhn, Thomas S. *The Copernican Revolution: Planetary Astronomy in the Development of Western Thought.* Cambridge, MA: Harvard University Press, 1957.

Manuel, Frank E. "Newton as Autocrat of Science." *Daedalus,* Summer 1968, pp. 969–1001.

Provine, William. "Evolution and the Foundations of Ethics." *MBL Science,* Winter 1988, v3n1, pp. 25–29. (Marine Biological Laboratory, Woods Hole, MA.)

Quarterly Review. "Giordano Bruno and Galileo Galilei." *Popular Science Monthly Supplement,* 1878, Volumes XIII–XX (bound volume S3), pp. 111–128.

Redondi, Pietro. *Galileo Heretic.* Princeton, NJ: Princeton University Press, 1987. (A novel reading of the Galileo/Urban affair, suggesting that the trial was a plea bargain designed to protect Galileo against even more serious charges of heresy for promoting the atomic theory of matter.)

Reston, James, Jr. *Galileo: A Life.* New York: HarperCollins, 1994.

Segre, Michael. *In the Wake of Galileo.* New Brunswick, NJ: Rutgers University Press, 1991.

Sharratt, Michael. *Galileo: Decisive Innovator.* Cambridge, England: Cambridge University Press, 1994.

2. Wallis versus Hobbes

Bold, Benjamin. *Famous Problems of Geometry and How to Solve Them.* New York: Dover Publications, 1982. (Reprint of 1969 edition, Van Nostrand Reinhold, slightly corrected.)

Boyer, Carl B. *The History of the Calculus and Its Conceptual Development.* New York: Dover Publications, 1959 (1949).

Chabot, Dana. "Thomas Hobbes: Skeptical Moralist." *American Political Science Review*, June 1995, pp. 401–410.

Cohen, I. Bernard. "Review of J. F. Scott, *The Mathematical Works of John Wallis, 1938.*" *Isis*, 1939, v30n3, pp. 529–532.

Dick, Oliver Lawson, editor. *Aubrey's Brief Lives*. Ann Arbor: University of Michigan Press, 1949/1957.

Eliot, P. F., editor. *French and English Philosophers: Descartes, Voltaire, Rousseau, Hobbes*. New York: P. F. Collier and Son, 1910 (*The Harvard Classics*, Vol. 34).

Gardner, Martin. "Mathematical Games: Incidental Information about the Extraordinary Number Pi." *Scientific American*, July 1960, pp. 154–156.

Hazard, Paul. *The European Mind, 1680–1715: The Critical Years*, reprint edition. New York: Fordham University Press, 1990.

Hinnant, Charles H. *Thomas Hobbes*. Boston: Twayne Publishers, 1977.

Hobbes, Thomas. *Leviathan*. New York: Penguin Books, 1986 (1651).

Malcolm, Noel, editor. *The Correspondence of Thomas Hobbes*, two volumes. New York: Oxford University Press, 1994.

Mintz, Samuel I. "Galileo, Hobbes, and the Circle of Perfection." *Isis*, July 1952, v43, pp. 98–100.

Mintz, Samuel I. "Hobbes." In C. Gillispie (editor), *Dictionary of Scientific Biography (DSB)*, Vol. 6, p. 449. New York: Scribner, 1972.

Mintz, Samuel I. *The Hunting of Leviathan: Seventeenth Century Reactions to the Materialism and Moral Philosophy of Thomas Hobbes*. Cambridge, England: Cambridge University Press, 1962.

Molesworth, Sir William, editor. *The English Works of Thomas Hobbes of Malmesbury*, 11 volumes. London: John Bohn (1839–1845) (reprinted 1962).

Robertson, George Croom. *Hobbes*. Edinburgh: William Blackwood & Sons, 1886.

Rogow, Arnold A. *Thomas Hobbes. Radical in the Service of Reaction*. New York: W. W. Norton, 1986.

Scott, J. F. *The Mathematical Works of John Wallis, D.D., F.R.S.* London: Taylor and Francis, 1938.

Shapin, Steven, and Schaffer, Simon. *Leviathan and the Air-Pump: Hobbes, Boyle, and the Experimental Life* (including a translation of Thomas Hobbes, *Dialogus Physicus de Natural Aeris*, by Simon Shaffer). Princeton, NJ: Princeton University Press, 1985.

Skinner, Quentin. "Bringing Back a New Hobbes," review of *The Correspondence of Thomas Hobbes* (edited by Noel Malcolm). *New York Review of Books*, April 4, 1996, v43n6, pp. 58–61 (unpaged online).

Smith, Preserved. *A History of Modern Culture: Vol. I. The Great Re-newal, 1543-1687* (1930); *Vol. II. The Enlightenment, 1687-1776* (1934). New York: Henry Holt. (Reprinted 1957 by Peter Smith.)

Watkins, J. W. N. *Hobbes's System of Ideas: A Study in the Political Significance of Philosophical Theories.* London: Hutchison University Library, 1965.

3. Newton versus Leibniz

Andrade, E.N. da C. *Sir Isaac Newton.* London: Collins, 1954.

Bell, E. T. *The Development of Mathematics,* second edition. New York: McGraw-Hill, 1945.

Berlinski, David. *A Tour of the Calculus.* New York: Pantheon Books, 1995.

Boyer, Carl B. *The History of the Calculus and Its Conceptual Development.* New York: Dover Publications, 1959 (1949).

Broad, William J. "Sir Isaac Newton: Mad as a Hatter." *Science,* September 18, 1981, v213, pp. 1341, 1342, 1344. Also letters, November 13, 1981, and March 5, 1982.

Bury, J. B. *The Idea of Progress.* New York: Dover Publications, 1960. (Reprint of original Macmillan edition, 1932, Chapter 19, "Progress in the Light of Evolution.")

Frankfurt, Harry G., editor. *Leibniz: A Collection of Critical Essays.* New York: Doubleday, 1972 (paperback). (Especially "Leibniz and Newton.")

Guillen, Michael. *Five Equations That Changed the World.* New York: Hyperion, 1995. (Especially sections on Newton, pp. 9-63; and the Bernoullis, pp. 65-117.)

Hall, A. Rupert. *From Galileo to Newton.* New York: Dover Publications, 1981 (Harper & Row, 1963).

Hall, A. Rupert. *Philosophers at War: The Quarrel Between Newton and Leibniz.* New York: Cambridge University Press, 1980.

Hall, A. Rupert, and Tilling, Laura, editors. *The Correspondence of Isaac Newton: Vol. 7, 1718-1727.* New York: Cambridge University Press, 1977.

Hathaway, Arthur S. "Further History of the Calculus." *Science,* February 13, 1920, pp. 166-167.

Hunt, Frederick Vinton. *Origins in Acoustics: The Science of Sound from Antiquity to the Age of Newton.* New Haven, CT: Yale University Press, 1978. (Especially the Newton-Leibniz feud, pp. 146ff.)

Latta, Robert, editor. *Leibniz: The Monadology and Other Philosophical Writings.* London: Oxford University Press, 1898.

Manuel, Frank E. "Newton as Autocrat of Science." *Daedalus,* Summer 1968, pp. 969–1001.

Merz, John Theodore. *Leibniz.* New York: Lippincott, 1884.

More, Louis T. *Isaac Newton: A Biography.* New York: Dover Publications, 1962 (1934).

Newton, Isaac. *Mathematical Principles of Natural Philosophy.* Chicago: Encyclopedia Britannica, 1955 (1687).

Peursen, C. A. van. *Leibniz.* New York: Dutton, 1970.

Price, Derek J. de Solla. *Little Science, Big Science.* New York: Columbia University Press, 1963. (Especially p. 68.)

Smith, Preserved. *A History of Modern Culture: Vol. I. The Great Renewal, 1543–1687* (1930); *Vol. II. The Enlightenment, 1687–1776* (1934). New York: Henry Holt. (Reprinted 1957 by Peter Smith.)

Spitz, L. W. "Leibniz's Significance for Historiography." *Isis,* 1952, v13, pp. 333–348.

Struik, Dirk J. *A Concise History of Mathematics.* New York: Dover Publications, 1967 (1948).

Westfall, Richard S. *Never At Rest: A Biography of Isaac Newton.* Cambridge, England: Cambridge University Press, 1980.

4. Voltaire versus Needham

Andrews, Wayne. *Voltaire.* New York: New Directions, 1981.

Besterman, Theodore, editor. *Voltaire.* New York: Harcourt, Brace & World, 1969.

Besterman, Theodore, editor. *The Works of Voltaire,* revised edition, with new translations by William F. Fleming. London: Blackwell, 1975 (original, Harcourt, Brace & World, 1969).

Bottiglia, William F. *Voltaire: A Collection of Critical Essays.* Englewood Cliffs, NJ: Prentice-Hall, 1968.

Brooks, Richard A., editor. *The Selected Letters of Voltaire.* New York: New York University Press, 1973.

Endore, Guy. *Voltaire! Voltaire!* New York: Simon & Schuster, 1961.

Gillespie, Charles S. "Voltaire." *Dictionary of Scientific Biography,* Vol. 14, pp. 83–85. New York: Scribner, 1976.

Glass, H. Bentley. "Maupertuis, a Forgotten Genius." *Scientific American,* October 1955, v193, pp. 100–110.

Haac, Oscar A. "Voltaire and Leibniz: Two Aspects of Rationalism." *Studies on Voltaire and the Eighteenth Century,* Vol. 25, pp. 795–809. Oxford, England: Voltaire Foundation at the Taylor Institution, 1963.

Mason, Haydn. *Voltaire.* New York: St. Martin's Press, 1975.

Meyer, Arthur William. *The Rise of Embryology.* Palo Alto, CA: Stanford University Press, 1939.

Needham, Joseph. *A History of Embryology,* second edition. New York: Abelard-Schuman, 1959 (1934).

Oppenheimer, Jane M. *Essays in the History of Embryology and Biology.* Cambridge, MA: MIT Press, 1967.

Orieux, Jean. *Voltaire.* Garden City, NY: Doubleday, 1979.

Perkins, Jean A. "Voltaire and the Natural Sciences." *Studies on Voltaire and the Eighteenth Century,* Vol. 37, pp. 61–76. Oxford, England: Voltaire Foundation at the Taylor Institution, 1965.

Prescott, F. "Spallanzani on Spontaneous Generation and Digestion." *Proceedings of the Royal Society of Medicine,* 1929–1930, v23, pp. 495–503.

Redman, Ben Ray, editor. *The Portable Voltaire.* New York: Viking Press, 1949.

Richter, P., and Ricardo, I. *Voltaire.* New York: Twayne, 1980.

Roe, Shirley A. *Matter, Life, and Generation.* Cambridge, England: Cambridge University Press, 1981.

Roe, Shirley A. "Voltaire versus Needham: Spontaneous Generation and the Nature of Miracles." Lecture at the New York Academy of Sciences, December 2, 1981.

Voltaire. *Candide and Other Stories.* New York: Alfred A. Knopf (Everyman's Library), 1992 (1759).

Voltaire. "A Dissertation by Dr. Akakia, Physician to the Pope" (1752). *The Works of Voltaire: A Contemporary Version,* translated by William F. Fleming, Vol. 19, Part 1, pp. 183–199. New York: St. Hubert Guild, E. R. Dumont, 1901.

Voltaire. *The Works of Voltaire: A Contemporary Version,* critique and biography by John Morley, translated by William F. Fleming (22 volumes). New York: St. Hubert Guild, E. R. Dumont, 1901 (1752).

Vulliamy, C. E. *Voltaire.* Port Washington, NY: Kennikat Press, 1970 (1930).

Westbrook, Rachel H. *John Turberville Needham and His Impact on the French Enlightenment.* Unpublished Ph.D. thesis, Columbia University, 1972.

5. Darwin's Bulldog versus Soapy Sam

Agassiz, Louis. "Prof. Agassiz on the Origin of Species." *American Journal of Science and Arts,* 1860, v79, pp. 142–154; reprinted in John C. Burnham (editor), *Science in America: Historical Selections.* New York: Holt, Rinehart and Winston, 1971.

Applebome, Peter. "Seventy Years after Scopes Trial, Creation Debate Lives." *New York Times,* March 10, 1996, pp. 1, 22.

Behe, Michael J. "Clueless at Oxford." *National Review,* October 14, 1996a, pp. 83–84. (Review of *Climbing Mount Improbable* by Richard Dawkins.)

Behe, Michael J. *Darwin's Black Box: The Biochemical Challenge to Evolution.* New York: The Free Press, 1996b.

Behe, Michael J. "Darwin under the Microscope." *New York Times,* October 29, 1996c, p. A25 (op ed.).

Benton, M.J. "Diversification and Extinction in the History of Life." *Science,* April 7, 1995, v268, pp. 52–67.

Berlinski, David. "The Deniable Darwin." *Commentary,* June 1996, pp. 19–29.

Berlinski, David. "The Soul of Man under Physics." *Commentary,* January 1996, pp. 38–46. (Berlinski's feelings about modern science.)

Berlinski, David. *A Tour of the Calculus.* New York: Pantheon Books, 1995.

Berreby, David. "Are Apes Naughty by Nature?" *New York Times Magazine,* January 26, 1997, pp. 38–39.

Berreby, David. "Enthralling or Exasperating: Select One." *New York Times,* September 24, 1996, pp. C1, C9.

Bishop, B. E. "Mendel's Opposition to Evolution and to Darwin." *Journal of Heredity,* May 1996, v87n3, pp. 205–213.

Boynton, Robert S. "The Birth of an Idea." *New Yorker,* October 7, 1996, pp. 72–81. (Darwin was no genius; why him?)

Brent, Peter. *Charles Darwin: A Man of Enlarged Curiosity.* New York: Harper & Row, 1981.

Bussey, Howard. "Chain of Being." *The Sciences,* March/April 1996, pp. 28–33. (Yeast.)

Campbell, Neil A. "A Conversation With John Maynard Smith." *American Biology Teacher,* October 1996, v59n7, pp. 408–412.

Caplan, Arthur. "What Controversy Tells Us about Science." *MBL Science,* Winter 1988, v3n1, pp. 20–24. (Marine Biological Laboratory, Woods Hole, MA.)

Caudill, Edward. "The Press and Tails of Darwin: Victorian Satire of Evolution." *Journalism History,* Autumn 1994, v20n3–4, pp. 107–115. (This is an excellent and entertaining paper and can be obtained online from the UMI Research I database.)

Clark, Ronald W. *The Survival of Charles Darwin: A Biography of a Man and an Idea.* New York: Random House, 1984.

Colp, Ralph, Jr. "I Will Gladly Do My Best: How Charles Darwin Obtained a Civil List Pension for Alfred Russel Wallace." *Isis,* 1992, v83, pp. 3–26.

Colson, Charles. "Planet of the Apes?" *Christianity Today,* August 12, 1996, v40n9, p. 64.

Cooper, Henry S. F. "Origins: The Backbone of Evolution." *Natural History,* June 1996, pp. 30–43.

Cravens, Hamilton. "The Evolution Controversy in America" (review of a book by the same name, by George E. Webb). *American Historical Review,* April 1996, v101n2, pp. 553–554.

Crook, Paul. *Darwinism, War and History: The Debate over the Biology of War from the 'Origin of Species' to the First World War.* Cambridge, England: Cambridge University Press, 1994.

Darwin, Charles. *The Origin of Species by Means of Natural Selection,* sixth edition, 1872 (1859); and *The Descent of Man and Selection in Relation to Sex* (1871) (combined edition). New York: The Modern Library, undated.

Darwin, Francis, editor. *The Autobiography of Charles Darwin and Selected Letters.* New York: Dover Publications, 1958 (1892).

Davidson, Eric H., et. al. "Origin of Belaterian Body Plans: Evolution of Developmental Regulatory Mechanisms." *Science,* November 24, 1996, v270, pp. 1319–1325.

Dawkins, Richard. *Climbing Mount Improbable.* New York: W. W. Norton, 1996.

de Camp, L. Sprague, and de Camp, Catherine Crook. *Darwin and His Great Discovery.* New York: Macmillan, 1972.

Degler, Carl N. "The Temptations of Evolutionary Ethics." *American Historical Review,* June 1996, v101n3, p. 838. (Review of *The Temptations of Evolutionary Ethics,* by Farber.)

Dennett, Daniel C. "Appraising Grace: What Evolutionary Good Is God?" *The Sciences,* January/February 1997, pp. 39–44. (Essay review of *Creation of the Sacred: Tracks of Biology in Early Religions,* by Walter Burkert.)

Dennett, Daniel C. *Darwin's Dangerous Idea: Evolution and the Meanings of Life.* New York: Simon & Schuster, 1995.

Desmond, Adrian, and Moore, James. *Darwin: The Life of a Tormented Evolutionist.* New York: Warner, 1991.

Eldridge, Niles. *Reinventing Darwin: The Great Debate at the High Table of Evolutionary Theory.* New York: John Wiley & Sons, 1995. (Eldridge and Stephen Jay Gould came up with the idea of punctuated equilibrium and ignited a furious debate about the true nature of evolution, involving geneticists vs. paleontologists.)

Farber, Paul Lawrence. *The Temptations of Evolutionary Ethics.* Berkeley: University of California Press, 1994.

Gatewood, Willard B. Essay review of *God's Own Scientists: Creationists in a Secular World,* by Christopher P. Toumey (Rutgers, 1994), and of *The Evolution Controversy in America,* by George E. Webb (University Press of Kentucky, 1994). *Isis,* 1995, v86n2, pp. 305–307.

Gillispie, Neil C. *Charles Darwin and the Problem of Creation*. Chicago: University of Chicago Press, 1979. (Especially Chapter 4, "Special Creation in the Origin: The Scientific Attack.")

Gould, Stephen Jay. *Dinosaur in a Haystack*. New York: Harmony Books/Crown Publishers, 1995. (Section on evolution, creationism.)

Gould, Stephen Jay. "Modified Grandeur." *Natural History*, March 1993, pp. 14–20. (A personal view of evolution and "grandeur.")

Gould, Stephen Jay. "The Tallest Tale." *Natural History*, May 1996, pp. 18ff. (The "neck of the giraffe [is] not a good example of Darwinian evolution.")

Grady, Wayne. "Darwin's American Pitbull." *Canadian Geographic*, March 1996, v116n2, p. 81 (online). (Review of Stephen Jay Gould's *Dinosaur in a Haystack*.)

Gray, Asa. "Review of Darwin's Theory on the Origin of Species by Means of Natural Selection." *American Journal of Science and Arts*, 1860, v79, pp. 153–184. Reprinted in John C. Burnham (editor), *Science in America: Historical Selections*. New York: Holt, Rinehart and Winston, 1971.

Haas, J. W., Jr. "The Biblical Flood: A Case Study of the Church's Response to Extrabiblical Evidence." *Theology Today*, October 1996, v53n3, pp. 401–404. (Review of the book by the same name, by David A. Young, Grand Rapids: Eerdmans, 1995.)

Hammond, Allen, and Margulis, Lynn. "Creationism as Science: Farewell to Newton, Einstein, Darwin. . . ." *Science 81*, December 1981, pp. 55–57.

Hitt, Jack. "On Earth As It Is in Heaven." *Harper's*, November 1996, v293 n1758, pp. 51–60. (Visit to the headquarters of creationist group.)

Holden, Constance. "Alabama Schools Disclaim Evolution." *Science*, November 24, 1995, p. 1305.

Holden, Constance. "The Vatican's Position Evolves." *Science*, November 1, 1996, v274n5288, p. 717.

Horgan, John. "Escaping in a Cloud of Ink." *Scientific American*, August 1995, pp. 37–41. (Profile of Stephen Jay Gould.)

Horgan, John. "The New Social Darwinists." *Scientific American*, October 1995, pp. 174–181.

Kerr, Richard A. "Geologists Debate Ancient Life and Fractured Crust: Embryos Give Clues to Early Evolution." *Science*, November 24, 1995, pp. 1300–1301.

Kimler, William. "Tracing Evolutionary Biology's Intellectual Phylogeny." *American Scientist*, March–April 1997, v85, pp. 177–178. (Review of *Life Splendid Drama: Evolutionary Biology and the Reconstruction of Life's Ancestry, 1860-1940*, by Peter J. Bowler, University of Chicago Press, 1996. Argues, in part, that historians of evolution have not paid enough attention to the work in biology that dominated the past century.)

Kohn, Marck. "Whigs and Hunters" (essay review of *River Out of Eden,* by Richard Dawkins, and *Reinventing Darwin,* by Niles Eldridge). *New Statesman & Society,* July 14, 1995, v8n361, pp. 34–35 (online).

Larson, Edward J. *Summer for the Gods: The Scopes Trial and America's Continuing Debate over Science and Religion.* New York: Basic Books, 1997.

Lewin, Roger. "Biology Is Not Postage Stamp Collecting." *Science,* May 14, 1982, v216, pp. 718–720. (Interview with Ernst Mayr.)

Lewin, Roger. *Bones of Contention: Controversies in the Search for Human Origins.* New York: Simon & Schuster, 1987.

Lewin, Roger. "Evolution's New Heretics." *Natural History,* May 1996, pp. 12–17.

Lewin, Roger. *Patterns in Evolution: The New Molecular View.* New York: Scientific American Library, 1997.

Livingstone, David N. *Darwin's Forgotten Defenders: The Encounter between Evangelical Theology and Evolutionary Thought.* Grand Rapids, MI: William B. Eerdmans Publishing, 1987.

Malik, Kenan. "The Beagle Sails Back into Fashion." *New Statesman,* December 6, 1996, pp. 35–36. (Social Darwinism.)

Margulis, Lynn, and Dolan, Michael F. "Swimming against the Current." *The Sciences,* January/February 1997, pp. 20–25. (Mergers of symbionts lead to large, functional evolutionary jumps; a possible evolutionary path to nucleated cells.)

Mayr, Ernst. *One Long Argument: Charles Darwin and the Genesis of Modern Evolutionary Thought.* Cambridge, MA: Harvard University Press, 1991.

McCollister, Betty. "Creation 'Science' vs. Religious Attitudes." *USA Today: The Magazine of the American Scene,* May 1996, v124n2612, pp. 74–76.

McDonald, Kim. "A Dispute over the Evolution of Birds." *Chronicle of Higher Education,* October 25, 1996, v43n9, pp. A14–A15.

Milner, Richard. "Charles Darwin and Associates, Ghostbusters." *Scientific American,* October 1996a, pp. 96–101.

Milner, Richard. *Charles Darwin: Evolution of a Naturalist.* New York: Facts on File, Inc., 1994.

Milner, Richard. *The Encyclopedia of Evolution. Humanity's Search for Its Origins.* New York: Facts on File, Inc., 1990.

Milner, Richard. "On What a Man Have I Been Wasting My Time" (review of *Charles Darwin's Letters: A Selection 1825–1859*). *Natural History,* May 1996b, pp. 6–7.

Murdoch, William W. "Theory for Biological Control: Recent Developments." *Ecology,* October 1996, v77n7, pp. 2001–2003.

Nesse, Randolph M., and Williams, George C. *Why We Get Sick: The New Science of Darwinian Medicine.* New York: Vintage Books, 1996. (Original hard cover, 1995.)

Numbers, Ronald L. "Creation Science." *Christian Century,* May 24, 1995, v112n18, pp. 574–575 (online).

Numbers, Ronald L. "Creationism in America." *Science,* November 5, 1982, v218, pp. 538–544.

Numbers, Ronald L. *The Creationists: The Evolution of Scientific Creation.* New York: Alfred A. Knopf, 1992.

Olroyd, D. R. *Darwinian Impacts.* Atlantic Highlands, NJ: Humanities Press, 1980.

Provine, William. "Evolution and the Foundation of Ethics." *MBL Science,* Winter 1988, v3n1, pp. 25–29. (Marine Biological Laboratory, Woods Hole, MA.)

Raloff, Janet. "When Science and Beliefs Collide." *Science News,* June 8, 1996, pp. 360–361.

Ramsay, M. A. "Darwinism, War and History. . . ." *Journal of Military History,* July 1996, v60n3, pp. 560–561.

Root-Bernstein, Robert S. "Darwin's Rib." *Discover Magazine,* September 1995, pp. 38–41.

Ryan, Michael. "Have Our Schools Heard the Wake-up Call?" *Parade Magazine,* January 19, 1997, pp. 8, 9.

Scott, Eugenie C. "Monkey Business." *The Sciences,* January/February 1996, pp. 20–25. (Follow-up responses, March/April, pp. 3ff.)

Shapiro, Robert. *Origins: A Skeptic's Guide to the Creation of Life on Earth.* New York: Summit Books, 1986.

Sholer, Jeffery L. "The Pope and Darwin." *US News & World Report,* November 4, 1996, v121n18, p. 12.

Shreeve, James. "Design for Living" (review of *Darwin's Black Box,* by Michael J. Behe). *New York Times Book Review,* August 4, 1996, p. 8.

Smith, Nancy F. "It's Just That Simple." *Audubon,* September 1996, v98n5, pp. 112–114. (Review of *Full House: The Spread of Excellence from Plato to Darwin,* by Stephen Jay Gould.)

Staff. "Biodiversity Is a Guarantee of Evolution: Interview with Werner Arber." *UNESCO Courier,* October 1996, n10, pp. 4–6.

Staff. "Denying Darwin: David Berlinsky and Critics." *Commentary,* September 1996, pp. 4–39.

Staff. "Evolution: The Dissent of Darwin." *Psychology Today,* January/February 1997, pp. 58–63. (Discussion between Richard Dawkins and Jaron Lanier.)

Stix, Gary. "Postdiluvian Science." *Scientific American,* January 1997, pp. 96–98.

Strahler, Arthur N. *Science and Earth History: The Evolution/Creation Controversy.* Buffalo, NY: Prometheus Books, 1987.

Tierney, Kevin. *Darrow: A Biography.* New York: Thomas Y. Crowell, Publishers, 1979. (Chapters 31 and 32, on the Scopes trial.)

Toulmin, Stephen, and Goodfield, June. *The Discovery of Time.* New York: Harper & Row, 1965. (Extensive section on the development of evolution and the objections to it, including those having to do with Kelvin; also, some background on the age-of-Earth controversy.)

Webb, George E. *The Evolution Controversy in America.* Lexington: University Press of Kentucky, 1994.

Wheeler, David L. "A Biochemist Urges Darwinists to Acknowledge the Role Played by an 'Intelligent Designer.'" *Chronicle of Higher Education,* November 1, 1996a, v43n10, pp. A13–A16.

Wheeler, David L. "An Eclectic Biologist Argues that Humans Are Not Evolution's Most Important Result; Bacteria Are." *Chronicle of Higher Education,* September 6, 1996b, v43n2, pp. A23–A24.

Wilford, John Noble. "Horses, Mollusks and the Evolution of Bigness." *New York Times,* January 21, 1997, pp. C1, C9.

Wilson, Edward O. *In Search of Nature.* Washington, DC: Island Press, 1996. (Seeking the origins of behavior.)

Wright, Robert. *The Moral Animal: Why We Are the Way We Are: The New Science of Evolutionary Psychology.* New York: Pantheon Books, 1994.

Wright, Robert. "Science and Original Sin: Evolutionary Biology Punctured the Notion of Six-Day Creation, but Biblical Themes of Good and Evil Are More Robust." *Time,* October 28, 1996, pp. 76–77. (Evolutionary psychology; morality.)

6. Lord Kelvin versus Geologists and Biologists

Basalla, George, editor. *Victorian Science.* New York: Doubleday, 1970 (paperback).

Broad, William J. "Bugs Shape Landscape, Make Gold." *New York Times,* October 15, 1996, pp. C1, C8.

Brush, Stephen G. "Kelvin in His Times" (essay review of *Energy and Empire,* by Smith and Wise). *Science,* May 18, 1990, pp. 875–877.

Burchfield, Joe D. *Lord Kelvin and the Age of the Earth.* New York: Science History Publications, 1975 (paper, 1990).

Casson, Herbert N. "Kelvin: His Amazing Life and Worldwide Influence." London: *The Efficiency Magazine,* undated (circa 1927), pp. 10–254.

Cowen, Ron. "Interplanetary Odyssey: Can a Rock Journeying from Mars to Earth Carry Life?" *Science News*, September 28, 1996, pp. 204–205.

Dalrymple, G. Brent. *The Age of the Earth*. Palo Alto, CA: Stanford University Press, 1991.

Dean, Dennis R. "The Age of the Earth Controversy: Beginnings to Hutton." *Annals of Science*, 1981, v38, pp. 435–456.

Frederickson, James K., and Tullis, C. Onstott. "Microbes Deep Inside the Earth." *Scientific American*, October 1996, pp. 68–73.

Huxley, Thomas Henry. "Geological Reform," (Huxley's answer to William Thomson's "On Geological Time") in *Transactions of the Geological Society of Glasgow: Vol 3. Lay Sermons, Addresses, and Reviews*. New York: Appleton, 1876.

Rudwick, Martin J. S. *The Great Devonian Controversy: The Shaping of Scientific Knowledge among Gentlemanly Specialists*. Chicago: University of Chicago Press, 1985. (Although actually about the dating of certain puzzling rock strata and fossils in the 1830s and 1840s, the book includes some background material on the catastrophism/uniformitarianism debate.)

Smith, Crosbie, and Wise, M. Norton. *Energy and Empire: A Biographical Study of Lord Kelvin*. Cambridge, England: Cambridge University Press, 1989.

Smith, Norman F. *Millions and Billions of Years Ago: Dating Our Earth and Its Life*. New York: Franklin Watts, 1993.

Twain, Mark. *Letters from the Earth*, edited by Bernard DeVoto. New York: Harper & Row, 1962 (1938).

7. Cope versus Marsh

Bakker, Robert T. *The Dinosaur Heresies: New Theories Unlocking the Mystery of the Dinosaurs and Their Extinction*. New York: William Morrow, 1986.

Bakker, Robert T. *Raptor Red*. New York: Bantam Books, 1995. (A fictional account of one year in the life of a dinosaur, based loosely on his heretical ideas; these are also spelled out in a nonfiction epilogue.)

Colbert, Edwin H. *Dinosaurs, An Illustrated History*. Maplewood, NJ: Hammond, 1983.

Colbert, Edwin H. *Little Dinosaurs of Ghost Ranch* (Coelophysis). New York: Columbia University Press, 1995.

Colbert, Edwin H. *Men and Dinosaurs: The Search in Field and Laboratory*. New York: E. P. Dutton, 1968.

DiChristina, Mariette. "The Dinosaur Hunter." *Popular Science,* September 1996, pp. 41–45.

Fortey, Richard. *Fossils: The Key to the Past.* New York: Van Nostrand Reinhold, 1982.

Gore, Rick. "Dinosaurs." *National Geographic,* January 1993, pp. 2–53.

Holmes, Thom. *Fossil Feud: The Rivalry of the First American Dinosaur Hunters.* Persippany, NJ: Julian Messner, 1998. (Suitable for young adults or adults; good illustrations.)

Kerr, Richard A. "K-T Boundary. New Way to Read the Record Suggests Abrupt Extinction." *Science,* November 22, 1996, v274, pp. 1303–1304.

Krishtalka, Leonard. *Dinosaur Plots and Other Intrigues in Natural History.* New York: Avon Books, 1989 (paperback).

Lakes, Arthur. *Discovering Dinosaurs in the Old West.* Washington, DC: Smithsonian Institution Press, 1997. (The journal of one of Marsh's field hands, edited by Michael F. Kohl and John S. McIntosh.)

Lanham, Url. *The Bone Hunters.* New York: Columbia University Press, 1973.

Morell, Virginia. "A Cold, Hard Look at Dinosaurs." *Discover,* December 1996, pp. 98–108.

Morell, Virginia. "The Origin of Birds: The Dinosaur Debate." *Audubon,* March/April 1997, pp. 36–45.

Munsart, Craig A., and Van Gundy, Karen Alonzi. *Primary Dinosaur Investigations: How We Know What We Know.* Englewood, CO: Teacher Ideas Press, 1995. (Intended as a teaching tool for students, the book provides fascinating background information for anyone interested in dinosaurs.)

Officer, Charles, and Page, Jake. *The Great Dinosaur Extinction Controversy.* New York: Helix (Addison-Wesley), 1996.

Ostrom, John H., and McIntosh, John S. *Marsh's Dinosaurs: The Collections from Como Bluff.* New Haven, CT: Yale University Press, 1966.

Padian, Kevin. "The Continuing Debate over Avian Origins." *American Scientist,* March–April 1997, v85, pp. 178–180. (Essay review of *The Origin and Evolution of Birds,* by Alan Feduccia, New Haven, CT: Yale University Press, 1996.)

Psihoyos, Louie, with John Knoebber. *Hunting Dinosaurs.* New York: Random House, 1994.

Riley, Matthew K. "O. C. Marsh: New York's Pioneer Fossil Hunter." *Conservationist,* 1993, v48n3, pp. 6–9.

Rudwick, Martin J. *The Great Devonian Controversy.* Chicago: University of Chicago Press, 1985.

Schuchert, Charles, and LeVene, Clara. *O. C. Marsh: Pioneer in Pale-ontology.* New Haven, CT: Yale University Press, 1940; New York: Arno Press, 1978.

Shor, Elizabeth Noble. *The Fossil Feud Between E. D. Cope and O. C. Marsh.* Hicksville, NY: Exposition Press, 1974.

Simpson, George Gaylord. *Fossils and the History of Life.* New York: Scientific American Library, 1983.

Spalding, David A. E. *Dinosaur Hunters: Eccentric Amateurs and Ob-sessed Professionals.* Rocklin, CA: Prima Publishing, 1993.

Wheeler, Walter H. "The Uintatheres and the Cope–Marsh War." *Sci-ence,* April 22, 1960, v131, pp. 1171–1176.

Wilford, John Noble. "A New Look at Dinosaurs." *New York Times Magazine,* February 7, 1982, pp. 22ff.

8. Wegener versus Everybody

Cowan, Ron. "Getting the Drift on Continental Shifts." *Science News,* February 12, 1994, p. 110.

Dalziel, Ian W. D. "Earth Before Pangea." *Scientific American,* January 1995, pp. 58–63.

Gohau, Gabriel. *A History of Geology.* New Brunswick, NJ: Rutgers University Press, 1990. (Chapters 15, 16, 17.)

Kerr, Richard A. "Earth's Surface May Move Itself." *Science,* September 1, 1995a, v269n5228, pp. 1214–1215.

Kerr, Richard A. "How Far Did the West Wander?" *Science,* May 5, 1995b, v268, pp. 635–637. (Discusses a current controversy be-tween geologists and geophysicists.)

Le Grand, H. E. *Drifting Continents and Shifting Theories.* Cambridge, England: Cambridge University Press, 1988.

Marvin, Ursula B. *Continental Drift: The Evolution of a Concept.* Wash-ington, DC: Smithsonian Institution Press, 1974.

Miller, Russell. *Continents in Collision.* Alexandria, VA: Time-Life Books, 1983. (Not completely up-to-date but contains excellent historical material, and beautiful illustrations.)

Monastersky, Richard. "Tibet Reveals Its Squishy Underbelly." *Science News,* December 7, 1996a, p. 356.

Monastersky, Richard. "Why Is the Pacific So Big? Look Down Deep." *Science News,* October 5, 1996b, p. 213.

Moores, Eldridge. "The Story of Earth." *Earth,* December 1996, pp. 30–33. (Plate tectonics.)

Nelson, K. Douglas. "Partially Molten Middle Crust beneath Southern Tibet: Synthesis of Project INDEPTH Results." *Science,* December 6, 1996, v274n5293, pp. 1684–1687.

Pool, Robert. "Plot Thickens in Earth's Inside Story." *New Scientist,* September 21, 1996, p. 19.

Romm, James. "A New Forerunner for Continental Drift." *Nature,* February 3, 1994, pp. 407–408.

Rossbacher, Lisa A. *Recent Revolutions in Geology.* New York: Franklin Watts, 1986.

Staff. "Did the Earth Ever Freeze Over?" *New Scientist,* July 30, 1994, p. 17.

Staff. "Two Plates Are Better Than One." *Science News,* August 19, 1995, p. 123.

Sullivan, Walter. *Continents in Motion: The New Earth Debate,* second edition. New York: McGraw-Hill, 1991.

Svitil, Kathy A. "The Mantle Moves Us." *Discover,* June 1996, p. 34.

Taylor, S. Ross. "The Evolution of Continental Crust." *Scientific American,* January 1996, pp. 76–81.

Thompson, Susan J. *A Chronology of Geological Thinking from Antiquity to 1899.* Metuchen, NJ: Scarecrow Press, 1988.

Van Andel, Tjeerd H. *New Views on an Old Planet: A History of Global Change,* second edition. Cambridge, England: Cambridge University Press, 1994.

Wegener, Alfred. *The Origin of Continents and Oceans.* New York: Dover Publications, 1966. (Translation of fourth edition, 1929.)

Windley, Brian F. *The Evolving Continents,* second edition. New York: John Wiley & Sons, 1984.

9. Johanson versus the Leakeys

Altmann, Jeanne. "Out of East Africa" (essay review of *Ancestral Passions,* by Morell). *Science,* November 24, 1995, v270, pp. 1381–1383.

Augereau, Jean-François. "New Views on the Origins of Man." *World Press Review,* August 1994, v41n8, p. 42.

Bower, B. "Oldest Fossil Ape May Be Human Ancestor." *Science News,* April 19, 1997, v151, p. 239.

Boynton, Graham. "Digging for Glory" (essay review of *Ancestral Passions,* by Morell). *Audubon,* January 1996, pp. 102, 105.

Clark, G. A., and Lindly, J. M. "Modern Human Origins in the Levant and Western Asia: The Fossil and Archeological Evidence." *American Anthropology,* 1989, v91, pp. 962–978.

Culotta, Elizabeth. "New Hominid Crowds the Field." *Science,* August 18, 1995, v269n5226, p. 918.

Current Biography Yearbook, 1995. "Richard Leakey." New York: H. W. Wilson, 1995, pp. 340–343.

da Silva, Wilson. "Human Origins Thrown into Doubt." *New Scientist,* March 29, 1997, p. 18.

Dorfman, Andrea, et al. "Not So Extinct After All." *Time,* December 23, 1996, pp. 68–69.

Economist. "Ancestral Passions: The Leakey Family and the Quest for Humankind's Beginnings." July 22, 1995, v336n7924, p. 83.

Economist. "Continental Drift." February 26, 1994, p. 87. (Doubts about African origin of humans.)

Economist. "Scientific Books: Origins." June 20, 1981, p. 113.

Economist. "Skulls and Numbskulls." November 21, 1992, v325n7786, p. 103.

Falk, Dean. "The Mother of Us All?" *Bioscience,* February 1995, v45n2, pp. 108–110. (Review of *Ancestors: In Search of Human Origins,* by Johanson, Johanson, and Edgar.)

Freeman, Karen. "More Recent Migration of Humans from Africa Is Seen in DNA Study." *New York Times,* June 4, 1996, p. 11.

Gibbons, Ann. "*Homo Erectus* in Java: A 250,000-Year Anachronism." *Science,* December 13, 1996, v274, pp. 1841–1842.

Golden, Frederick. "First Lady of Fossils, Mary Nicol Leakey: 1913–1996." *Time,* December 23, 1996, p. 69.

Gore, Rick. "Expanding Worlds." *National Geographic,* May 1997a, pp. 84–109.

Gore, Rick. "The First Steps." *National Geographic,* February 1997b, pp. 72–99. (Ongoing series, "The Dawn of Humans.")

Gorman, Christine. "On Its Own Two Feet." *Time,* August 28, 1995, pp. 58–60.

Johanson, Donald C. "Face-to-Face with Lucy's Family." *National Geographic,* March 1996, pp. 96–117.

Johanson, Donald. "A Skull to Chew On." *Natural History,* May 1993, pp. 52, 53. (The Black Skull, KNM-WT 17000, and parallel evolution.)

Johanson, Donald C., and Blake, Edgar. *From Lucy to Language.* New York: Simon & Schuster, 1996.

Johanson, Donald, and Edey, Maitland. *Lucy: The Beginnings of Humankind.* New York: Simon & Schuster, 1981.

Johanson, Donald, Johanson, Lenora, and Edgar, Blake. *Ancestors: In Search of Human Origins.* New York: Villard Books, 1994.

Johanson, Donald, and Shreeve, James. *Lucy's Child: The Discovery of a Human Ancestor.* New York: Morrow, 1989.

Johanson, Donald, and White, Tim D. "A Systematic Assessment of Early African Hominids." *Science,* 1979, v202, pp. 321–330.

Kern, Edward P. H. "Battle of the Bones: A Fresh Dispute over the Origins of Man." *Life,* December 1981, pp. 109–120.

Kluger, Jeffrey. "Not So Extinct after All." *Time*, December 23, 1996, pp. 68, 69.

Larick, Roy, and Ciochon, Russell L. "The African Emergence and Early Asian Dispersals of the Genus Homo." *American Scientist*, November–December 1996, v84, pp. 538–551.

Leakey, Mary. *Disclosing the Past: An Autobiography*. Garden City, NY: Doubleday, 1984.

Leakey, Meave. "The Farthest Horizon." *National Geographic*, September 1995, pp. 38–51.

Leakey, Richard. *"Homo Erectus Unearthed (A Fossil Skeleton 1,600,000 Years Old)." National Geographic*, November 1985, pp. 624–629.

Leakey, Richard. *The Making of Mankind*. New York: E. P. Dutton, 1981.

Leakey, Richard. *One Life*. Salem, MA: Salem House, 1984.

Leakey, Richard. *The Origin of Mankind*. New York: Basic Books, 1994.

Leakey, Richard E., and Lewin, Roger. *Origins: In Search of What Makes Us Human*. New York: E. P. Dutton, 1977.

Leakey, Richard, and Lewin, Roger. *Origins Reconsidered: In Search of What Makes Us Human*. New York: Doubleday, 1992.

Lemonick, Michael D. "Picks & Pans: Ancestral Passions." *People Weekly*, October 2, 1995, v44n14, pp. 32, 34.

Lewin, Roger. "Bones of Contention." *New Scientist*, November 4, 1995, pp. 14, 15.

Lewin, Roger. *Bones of Contention: Controversies in the Search for Human Origins*. New York: Simon & Schuster, 1987.

Lewin, Roger. "Family Feuds." *New Scientist*, January 24, 1988, pp. 36–40.

Lovejoy, C. Owen. "The Origin of Man." *Science*, January 23, 1981, pp. 341–350.

Maddox, Brenda. "Hominid Dreams" (essay review of *Ancestral Passions*, by Morell). *New York Times Book Review*, August 6, 1995, p. 28.

Major, John S. "The Secret of 'Leakey Luck.'" *Time*, August 28, 1995, p. 60.

McAuliffe, Sharon. "Lucy's Father." *Omni*, May 1994, pp. 34–39, 80, 83–86.

Menon, Shanti. "Neanderthal Noses." *Discover*, March 1997, p. 30.

Morell, Virginia. *Ancestral Passions: The Leakey Family and the Quest for Humankind's Beginnings*. New York: Simon & Schuster, 1995.

Morell, Virgina. "The Most Dangerous Game." *New York Times Magazine*, January 7, 1996, p. 23.

New York Times. "Richard Leakey: The Challenger in Dispute on Human Evolution." February 18, 1979, p. 41.

Nichols, Mark. "The Origins of Man." *Maclean's,* December 23, 1996, v109n52, p. 69.

Pieg, Pascal, and Verrechia, Nicole. *Lucy and Her Times.* New York: Henry Holt, 1996. (A lighthearted look at the "primeval world.")

Pope, Gregory G. "Ancient Asia's Cutting Edge." *Natural History,* May 1993, pp. 54–59. (Tool evidence in China.)

Rennie, John. "Fossils of Early Man: The Finds and the News." *New York Times,* June 25, 1996, pp. C1, C9.

Rensberger, Boyce. "Rival Anthropologists Divide on 'Pre-Human' Find." *New York Times,* February 18, 1979, pp. 1, 41.

"Roots" ("Human Origins, 1994," roundup). *Discover,* January 1995, pp. 37–42.

Shreeve, James. "'Lucy,' Crucial Early Human Ancestor, Finally Gets a Head." *Science,* April 1, 1994, v264, pp. 34–35.

Shreeve, James. "Sexing Fossils: A Boy Named Lucy?" *Science,* November 24, 1995, v270, pp. 1297–1298.

Shreeve, James. "Sunset on the Savanna." *Discover,* July 1996, pp. 116–125.

Tattersall, Ian. "Out of Africa Again . . . and Again?" *Scientific American,* April 1997, pp. 60–67.

Vrba, Elisabeth S. "The Pulse That Produced Us." *Natural History,* March 1993, pp. 47–51. (Antelopes and early humans.)

Walker, Alan, and Shipman, Pat. *The Wisdom of the Bones: In Search of Human Origins.* New York: Alfred A. Knopf, 1996.

Weaver, Kenneth F. "The Search for Our Ancestors." *National Geographic,* November 1985, pp. 560–623.

Wilford, John Noble. "Ancient German Spears Tell of Mighty Hunters of Stone Age." *New York Times,* March 4, 1997a, p. C6.

Wilford, John Noble. "The Leakeys: A Towering Reputation." *New York Times,* October 30, 1984, pp. C1, C9.

Wilford, John Noble. "The New Leader of a Fossil-Hunting Dynasty." *New York Times,* November 7, 1995, pp. C1, C6.

Wilford, John Noble. "Not About Eve." *New York Times Book Review,* February 2, 1997b, p. 19.

Wilford, John Noble. "Three Human Species Coexisted Eons Ago, New Data Suggest." *New York Times,* December 13, 1996a, pp. 1, B14.

Wilford, John Noble. "2.3-Million-Year-Old Jaw Extends Human Family." *New York Times,* November 19, 1996b, pp. 1, C5.

Wilford, John Noble. "Which Came First, Tall or Smart?" *New York Times Book Review,* December 1, 1996c, p. 7. (Review of *From Lucy to Language,* by Johanson and Edgar.)

Willis, Delta. *The Leakey Family: Leaders in the Search for Human Origins*. New York: Facts on File, 1992.

10. Derek Freeman versus Margaret Mead

Brady, Ivan. "The Samoa Reader: Last Word or Lost Horizon?" (review of *The Samoa Reader*, by Hiram Caton). *Current Anthropology*, August–October 1991, v32n4, pp. 497–500.

Caton, Hiram. *The Samoa Reader: Anthropologists Take Stock*. Lanham, MD: University Press of America, 1990.

Côté, James E. *Adolescent Storm and Stress: An Evaluation of the Mead–Freeman Controversy*. Hillsdale, NJ: Lawrence Erlbaum, 1994.

Côté, James E. "Much Ado about Nothing: The 'Fateful Hoaxing' of Margaret Mead." *Skeptical Inquirer*, November/December 1998, v22n6, pp. 29–34.

Côté, James E. "Was Mead Wrong About Coming of Age in Samoa? An Analysis of the Mead/Freeman Controversy for Scholars of Adolescence and Human Development." *Journal of Youth and Adolescence*, 1992, v21n5, pp. 499–527.

di Leonardo, Micaela. "Patterns of Culture Wars." *Nation*, April 8, 1996, v262n14, pp. 25–29.

Freeman, Derek. "Fa'apua'a Fa'amu and Margaret Mead." *American Anthropologist*, December 1989, v91n4, pp. 1017–1022.

Freeman, Derek. *The Fateful Hoaxing of Margaret Mead: A Historical Analysis of Her Samoan Research*. Boulder, CO: Westview, 1998.

Freeman, Derek. *Margaret Mead and the Heretic: The Making and Unmaking of an Anthropological Myth*. Ringwood (Australia): Penguin Books, 1996.

Freeman, Derek. *Margaret Mead and Samoa: The Making and Unmaking of an Anthropological Myth*. Cambridge, MA: Harvard University Press, 1983.

Freeman, Derek. "On Franz Boas and the Samoan Researches of Margaret Mead." *Current Anthropology*, June 1991, v32n3, pp. 322–330.

Freeman, Derek. "There's Tricks I' the World: An Historical Analysis of the Samoan Researches of Margaret Mead." *Visual Anthropology Review*, Spring 1991, v7, pp. 103–128.

Goodman, R. A. *Mead's Coming of Age in Samoa: A Dissenting View*. Oakland, CA: Pipperline Press, 1983.

Grant, Nicole J. "From Margaret Mead's Field Notes: What Counted as 'Sex' in Samoa?" *American Anthropologist*, December 1995, v97n4, pp. 678–682.

Harris, Marvin. "Margaret and the Giant Killer." *The Sciences*, July–August 1983, v23, pp. 18–21.

Holmes, Lowell D. *Quest for the Real Samoa: The Mead/Freeman Controversy & Beyond*. South Hadley, MA: Bergin & Garvey Publishers, 1987.

Holmes, Lowell D. *A Restudy of Manu'an Culture: A Problem in Methodology*. Ph.D. dissertation, Northwestern University, 1957.

Holmes, Lowell D. "South Seas Squall: Derek Freeman's Long-Nurtured, Ill-Natured Attack on Margaret Mead." *The Sciences*, 1983, v23, pp. 14–18.

Howard, Jane. *Margaret Mead: A Life*. New York: Simon & Schuster, 1984.

Kempermann, Gerd, Kuhn, H. George, and Gage, Fred H. "More Hippocampal Neurons in Adult Mice Living in Any Enriched Environment." *Nature*, April 3, 1997, v386n6624, pp. 493–495.

McDowell, Edwin. "New Samoa Book Challenges Margaret Mead's Conclusions." *New York Times*, January 31, 1983, pp. 1, C21.

Mead, Margaret. *Coming of Age in Samoa: A Psychological Study of Primitive Youth for Western Civilization*. New York: American Museum of Natural History, 1973.

Monaghan, Peter. "Research on Samoan Life Finds New Backing for His Claims." *Chronicle of Higher Education*, August 2, 1989, pp. A5, A6.

Muuss, R. E. *Theories of Adolescence*, fifth edition. New York: Random House, 1988.

Orans, Martin. "Mead Misrepresented." *Science*, March 12, 1999, v283, pp. 164–165.

Orans, Martin. *Not Even Wrong: Margaret Mead, Derek Freeman, and the Samoans*. Novato, CA: Chandler and Sharp Publishers, 1996.

Rensberger, Boyce. "The Nature–Nurture Debate: Two Portraits." *Science 83*, April, 1983, v4n3. (1. Margaret Mead, pp. 28–37; 2. On Becoming Human [Edward O. Wilson], pp. 38–46.)

Rensberger, Boyce. "A Pioneer and an Innovator." *New York Times*, November 16, 1978, pp. 1, D18.

Rubin, Vera. "Margaret Mead and Samoa: The Making and Unmaking of an Anthropological Myth" (review). *American Journal of Orthopsychiatry*, July 1983, v53n3, pp. 550–554.

Shankman, Paul. "Margaret Mead, Derek Freeman, and the Issue of Evolution." *Skeptical Inquirer*, November/December 1998, v22n6, pp. 35–39.

Whitman, Alden. "Margaret Mead Is Dead of Cancer at 76." *New York Times*, November 16, 1978, pp. 1, D18.

Wilson, Edward O. *Sociobiology: The New Synthesis*. Cambridge, MA: Harvard University Press, 1975.

INDEX

Wilson, Edward O., 181
Wolff, Kaspar Friedrich, 76
work, heat and, 110–11

X rays, 92

Yale Peabody Museum, 125

Yale University, 121, 125, 127, 138
Yukawa, Hideki, 23

*Zinjanthropus boisei. See Australo-
pithecus boisei*